GLYCOME INFORMATICS

METHODS AND APPLICATIONS

CHAPMAN & HALL/CRC
Mathematical and Computational Biology Series

Aims and scope:
This series aims to capture new developments and summarize what is known over the whole spectrum of mathematical and computational biology and medicine. It seeks to encourage the integration of mathematical, statistical and computational methods into biology by publishing a broad range of textbooks, reference works and handbooks. The titles included in the series are meant to appeal to students, researchers and professionals in the mathematical, statistical and computational sciences, fundamental biology and bioengineering, as well as interdisciplinary researchers involved in the field. The inclusion of concrete examples and applications, and programming techniques and examples, is highly encouraged.

Series Editors

Alison M. Etheridge
Department of Statistics
University of Oxford

Louis J. Gross
Department of Ecology and Evolutionary Biology
University of Tennessee

Suzanne Lenhart
Department of Mathematics
University of Tennessee

Philip K. Maini
Mathematical Institute
University of Oxford

Shoba Ranganathan
Research Institute of Biotechnology
Macquarie University

Hershel M. Safer
Weizmann Institute of Science
Bioinformatics & Bio Computing

Eberhard O. Voit
The Wallace H. Couter Department of Biomedical Engineering
Georgia Tech and Emory University

Proposals for the series should be submitted to one of the series editors above or directly to:
CRC Press, Taylor & Francis Group
4th, Floor, Albert House
1-4 Singer Street
London EC2A 4BQ
UK

Published Titles

Bioinformatics: A Practical Approach
Shui Qing Ye

Cancer Modelling and Simulation
Luigi Preziosi

**Combinatorial Pattern Matching
Algorithms in Computational Biology
Using Perl and R**
Gabriel Valiente

**Computational Biology: A Statistical
Mechanics Perspective**
Ralf Blossey

**Computational Neuroscience: A
Comprehensive Approach**
Jianfeng Feng

**Data Analysis Tools for DNA
Microarrays**
Sorin Draghici

**Differential Equations and
Mathematical Biology**
D.S. Jones and B.D. Sleeman

Engineering Genetic Circuits
Chris J. Myers

**Exactly Solvable Models of Biological
Invasion**
Sergei V. Petrovskii and Bai-Lian Li

**Gene Expression Studies Using
Affymetrix Microarrays**
Hinrich Göhlmann and Willem Talloen

**Glycome Informatics: Methods and
Applications**
Kiyoko F. Aoki-Kinoshita

**Handbook of Hidden Markov Models in
Bioinformatics**
Martin Gollery

Introduction to Bioinformatics
Anna Tramontano

**An Introduction to Systems Biology:
Design Principles of Biological Circuits**
Uri Alon

Kinetic Modelling in Systems Biology
Oleg Demin and Igor Goryanin

Knowledge Discovery in Proteomics
Igor Jurisica and Dennis Wigle

**Meta-analysis and Combining
Information in Genetics and Genomics**
Rudy Guerra and Darlene R. Goldstein

**Modeling and Simulation of Capsules
and Biological Cells**
C. Pozrikidis

**Niche Modeling: Predictions from
Statistical Distributions**
David Stockwell

**Normal Mode Analysis: Theory and
Applications to Biological and Chemical
Systems**
Qiang Cui and Ivet Bahar

**Optimal Control Applied to Biological
Models**
Suzanne Lenhart and John T. Workman

**Pattern Discovery in Bioinformatics:
Theory & Algorithms**
Laxmi Parida

Python for Bioinformatics
Sebastian Bassi

Spatial Ecology
*Stephen Cantrell, Chris Cosner, and
Shigui Ruan*

**Spatiotemporal Patterns in Ecology
and Epidemiology: Theory, Models, and
Simulation**
*Horst Malchow, Sergei V. Petrovskii, and
Ezio Venturino*

**Stochastic Modelling for Systems
Biology**
Darren J. Wilkinson

**Structural Bioinformatics: An
Algorithmic Approach**
Forbes J. Burkowski

**The Ten Most Wanted Solutions in
Protein Bioinformatics**
Anna Tramontano

Chapman & Hall/CRC Mathematical and Computational Biology Series

GLYCOME INFORMATICS

METHODS AND APPLICATIONS

KIYOKO F. AOKI-KINOSHITA

CRC Press
Taylor & Francis Group
Boca Raton London New York

CRC Press is an imprint of the
Taylor & Francis Group, an **informa** business

A CHAPMAN & HALL BOOK

First published 2010 by Chapman & Hall

Published 2019 by CRC Press
Taylor & Francis Group
6000 Broken Sound Parkway NW, Suite 300
Boca Raton, FL 33487-2742

First issued in paperback 2019

No claim to original U.S. Government works

ISBN-13: 978-0-367-45243-8 (pbk)
ISBN-13: 978-1-4200-8334-7 (hbk)

Visit the Taylor & Francis Web site at
http://www.taylorandfrancis.com

and the CRC Press Web site at
http://www.crcpress.com

Library of Congress Cataloging-in-Publication Data

Aoki-Kinoshita, Kiyoko F.
   Glycome informatics : methods and applications / Kiyoko F. Aoki Kinoshita.
      p. cm. -- (CHAPMAN & HALL/CRC mathematical and computational biology
      series ; 28)
   Includes bibliographical references and index.
   ISBN 978-1-4200-8334-7 (alk. paper)
   1. Glycoconjugates. 2. Glycomics. 3. Bioinformatics. I. Title.

QP702.G577A55 2009
572'.567--dc22   2009017459

Contents

List of Tables

List of Figures

About the Author

Kiyoko F. Aoki-Kinoshita simultaneously received her bachelor's and master's degrees of science in computer science from Northwestern University in 1996, after which she received her doctorate in computer engineering from Northwestern in 1999 under Dr. D. T. Lee. She was employed at BioDiscovery, Inc. in Los Angeles, California as a senior software engineer before moving to Kyoto, Japan, to work as a post-doctoral researcher at the Bioinformatics Center, Institute of Chemical Research, Kyoto University, under Drs. Hiroshi Mamitsuka and Minoru Kanehisa. There, she developed various algorithmic and data mining methods for analyzing the glycan structure data that were accumulated in the KEGG GLYCAN database.

Since then, she has joined the faculty in the Department of Bioinformatics, Faculty of Engineering, Soka University, in Tokyo, Japan and is now an associate professor teaching bioinformatics. She is also involved in several research projects pertaining to the understanding of glycan function based on their structure as well as the recognition patterns of glycan structures by other proteins and even viruses. She has also begun developing a Web resource called RINGS (Resource for INformatics of Glycomes at Soka) that is still in its infancy, but is intended to freely provide many of the informatics algorithms and methods described in this book over the Web such that scientists may utilize them easily.

Chapter 1

Introduction to Glycobiology

Welcome to the field of glycome informatics! This is the first book to cover all known informatics methods pertaining to the study of glycans, or glycobiology. We will begin with an introduction to the field of glycobiology, upon which these methods have been developed. Because it is impossible to introduce the entire field of glycobiology in one chapter, an introduction to the basic knowledge required to understand this book is provided. For more detailed information, readers are referred to a recently published comprehensive guide to the glycosciences by Kamerling et al. (2007).

Glycobiology is the field of research pertaining to the study of the structure, biosynthesis and biology (function) of glycans, which include glycosides such as oligosaccharides, polysaccharides, and glycoconjugates. Why is glycobiology important? It is known now that glycobiology is relevant to every living species, with glycans decorating the cells of all species. Considering that the human genome, for example, contains a much smaller number of genes than was originally expected, glycans and lipids provide an answer as to how such a complex system as the human body can function. Lipids and carbohydrates[1] can serve as intermediates in generating energy and as signaling effectors, recognition markers, and structural components. Especially in cell-cell, cell-matrix, and cell-molecule interactions, carbohydrates are particularly important for complex multicellular organisms. Thus the understanding of how these molecules function may be a major key in understanding the genome and the biological processes of complex systems (Varki et al. (2008)).

1.1 Roles of carbohydrates

Glycans may be analogized to access card readers to cells by which proteins function as cards containing access codes. Glycans decorate the surface of cells either directly or on cell surface proteins such that other biomolecules such as viruses, bacteria, pathogens and other proteins may recognize the

---

[1] The term carbohydrate, oligosaccharides, sugar chains and glycan will be used interchangeably throughout this book.

1

appropriate structures to bind to the target cells. The difficulty is that the glycans are flexible, and that depending on the environment, the structures of the glycans change. Using the analogy above, these access card readers may change their codes according to the temperature, humidity, or even time of day, for example. Thus proteins are allowed access only when the right conditions are met. In biological terms, the functions of many proteins are affected by glycan structure conformations which are governed by environmental conditions, thus emphasizing the importance of glycans and their functions on proteins.

In general, there are two broad categories into which the biological roles of glycans may be divided: (1) the structural and modulatory properties of glycans and (2) the specific recognition of glycans by other molecules, which include glycan-binding proteins (GBPs). There are two types of GBPs: intrinsic and extrinsic; the former referring to those that recognize glycans within the same organism, and the latter referring to those that recognize glycans from a different organism. Intrinsic GBPs generally function to mediate cell-cell interactions or to recognize extracellular molecules; they may also recognize glycans on the same cell. In contrast, extrinsic GBPs include pathogenic microbes, toxins and molecules mediating symbiotic relationships. These contrary roles of glycan recognition are suggested to act as opposing selective forces, which consequently affect evolutionary changes in biological systems (Varki et al. (2008)).

1.2   Glycan structures

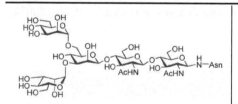

FIGURE 1.1:   The chemical structure of the *N*-glycan core structure.

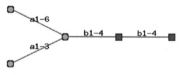

FIGURE 1.2:   The N-glycan core structure represented with monosaccharide symbols.

Complex carbohydrates are composed of monosaccharides that are covalently linked by glycosidic bonds, either in the α or β forms. Monosaccharides have the general empirical formula $C_x(H20)_n$, where $n = \{3, 4, \ldots, 9\}$, usually forming a ring of carbon atoms numbered from C1, as in Figure 1.3. The overall configuration (D or L) of each monosaccharide is determined by the absolute configuration of the stereogenic center furthest from the highest numbered asymmetric carbon atom, which is C5 in hexoses and C4 in pentoses. Figure 1.4 illustrates the D and L configurations of glucose. Alternatively, the chiral center may be labeled as R or S, depending on the annotator. Thus some databases will allow the definition of glycans using either the D/L or R/S systems.

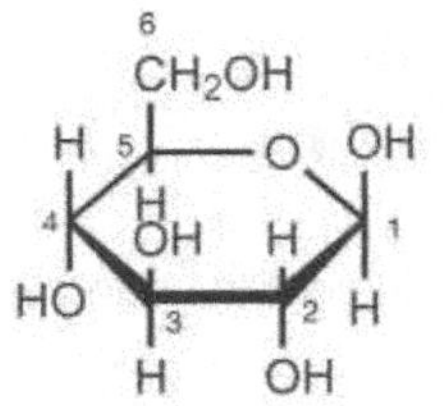

FIGURE 1.3: Glucose monosaccharide with the carbon numbers numbered from C1 through C6.

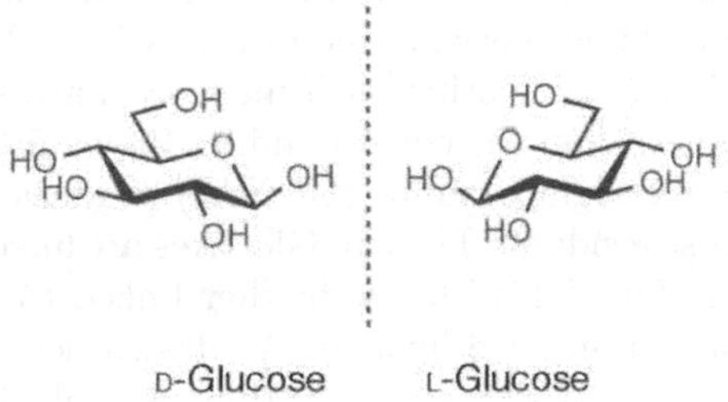

D-Glucose   L-Glucose

FIGURE 1.4: D and L configurations of glucose.

Monosaccharides most commonly form five- or six-membered rings due to chemical stability. A five-membered ring is called a heptose (or furanose), and a six-membered ring is called a hexose (or pyranose). When cyclized into these rings, monosaccharides acquire an asymmetric center called the anomeric carbon (the carbon atom having hemiacetal functionality, such as the C1 in the ring form of glucose as in Figure 1.3 or the C2 for sialic acids). Two stereoisomers can be formed because the anomeric hydroxy group can assume two possible orientations. When the anomeric carbon and the stereogenic center furthest away from it are the same, the monosaccharide is defined as the α anomer; when they are different, it is defined as the β anomer. Figure 1.5 illustrates how the anomeric configuration is determined. These anomeric configurations may change when monosaccharides are joined to one another, as described in Section 1.4.

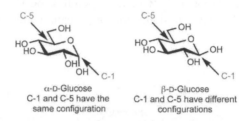

α-D-Glucose
C-1 and C-5 have the
same configuration

β-D-Glucose
C-1 and C-5 have different
configurations

FIGURE 1.5: The anomeric center is determined by the configuration of opposing carbon atoms.

Two monosaccharide units are joined together by a glycosidic bond, which is formed between the anomeric carbon of one monosaccharide to a hydroxy group of another. Unlike DNA and amino acid sequences, however, monosaccharides may be linked to more than one other monosaccharide, such that they form branched, tree structures. For example, the chemical structure of the N-linked glycan core structure is given in Figure 1.1. This core structure in fact consists of five building blocks of monosaccharides. Thus glycans can be displayed with symbols representing each monosaccharide connected by glycosidic linkages as lines, as in the corresponding Figure 1.2. The dark shaded squares represent N-acetylglucosamine (GlcNAc) residues and the shaded circles represent mannose residues. The two GlcNAcs are linked in a β1-4 linkage, as well as the first βMan. This Man is further linked to two other mannose residues by an α1-3 and an α1-6 linkage. As drawn here, carbohydrates are most classically drawn as a tree in a two-dimensional plane, with the root monosaccharide placed at the right-most position, called the reducing end of the glycan, and children branching out towards the left, referred to as the non-reducing end. Each node represents a monosaccharide, and each edge represents a glycosidic linkage, which includes the carbon numbers that are linked and the anomeric conformation. Note, however, that some cyclic glycans also exist. These types of glycans are rare and not considered in this book.

As mentioned earlier, glycan structures are flexible and may change their conformations depending on environmental conditions. Thus there is much research in analyzing glycan structures in three-dimensional (3D) space. In

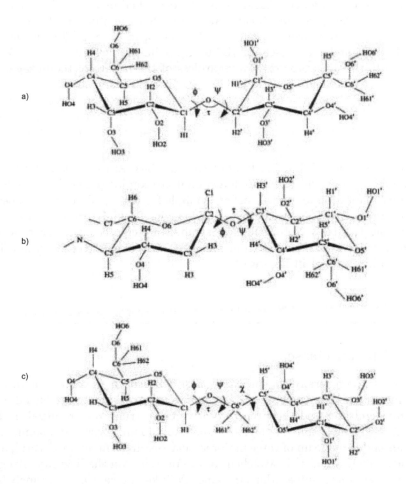

FIGURE 1.6: Torsion angles of glycosidic linkages. The atoms of the reducing end monosaccharide are labeled with primes (e.g. O1′, C1′). In general, (a) Φ is defined as H1-C1-O-Cx' and Ψ as C1-O-Cx'-Hx' for a 1-x linkage, where $x = \{1, 2, 3, 4\}$. O refers to the glycosidic oxygen regardless of the nature of the linkage. In the case of sialic acid, (b) Φ is defined as C1-C2-O-Cx', and in the case of 1-6 linkages, (c) Φ is H1-C1-O-C6′, Ψ is C1-O-C6′-C5′, and a third torsion angle χ is defined as O-C6′-C5′-H5′. The bond angle τ is defined as C1-O-C′. *Reused by permission of CRC press.*

fact, many of the databases described in Chapter 3 contain three-dimensional data of glycan structures. In 3D space, the relative orientation of the two monosaccharides in a disaccharide is usually described by the torsion angles Φ and Ψ around the glycosidic bonds. Assuming that the atoms of the reducing end are labeled with primes (e.g. O1′, C1′, etc.), in general, Φ is defined as H1-C1-O-Cx' and Ψ as C1-O-Cx'-Hx' for a 1-x linkage, where $x = \{1, 2, 3, 4\}$. O refers to the glycosidic oxygen regardless of the nature of the linkage. In the case of sialic acid, Φ is defined as C1-C2-O-Cx', and in the case of 1-6 linkages, Φ is H1-C1-O-C6′, Ψ is C1-O-C6′-C5′, and a third torsion angle χ (sometimes denoted as ω) is defined as O-C6′-C5′-H5′. The bond angle τ is defined as C1-O-C′. Figure 1.6 illustrates these torsion angles (Rao et al. (1998)).

Since the basic building blocks of glycans are basically consistent for mammalian cells, the Consortium for Functional Glycomics (CFG) has proposed a standard notation for glycans using symbols for commonly found monosaccharide residues. This standard is provided in Table 1.1 and will be used throughout this text. Note that these symbols are colored such that they are distinguishable even if printed in monochrome. An example of a glycan using this notation is given in Figure 1.2.

1.3　Glycan classes

Glycans are generally classified into classes based on the core structure of the glycan, which is composed of the monosaccharides at the reducing end. The major classes and their representative core structures are listed in Table 1.2. *N*-glycans comprise the most commonly found glycans in mammalian systems and is made up of three major subclasses: high-mannose, hybrid, and complex. They are usually attached to core proteins via the R-group nitrogen (N) of asparagine residues, thus its name. *N*-linked glycans are extremely important for proper folding of proteins in eukaryotic cells. As described later, chaperone proteins participate in *N*-glycan biosynthesis in order to ensure that proteins are properly folded. Steric effects of *N*-glycans also contribute to protein folding by blocking cysteine residues such that disulfide bonds are not formed, for example. *N*-glycans also play various roles in cell-cell interactions and protein targetting, but oftentimes it is not the core structure that is recognized, but the terminal structures, which may also be found on other types of glycans.

O-glycans are relatively smaller structures, usually attached to core proteins via serine or threonine residues. Mucins are heavily *O*-glycosylated glycoproteins, containing VNTR (variable number of tandem repeat) regions that are rich in Ser/Thr acceptor sites of *O*-glycans. Other amino acids that may be

TABLE 1.1: The standard representation of monosaccharides as proposed by the CFG. These are colored such that the symbols are distinguishable even if printed in monochrome.

Symbol	Abbreviation	Name
▲	Fuc	fucose
◯	Gal	galactose
◸	GalN	galactosamine
◈	GalA	galacturonic acid
☐	GalNAc	N-acetylgalactosamine
●	Glc	glucose
◣	GlcN	glucosamine
◇	GlcA	glucuronic acid
■	GlcNAc	N-acetylglucosamine
◈	IdoA	iduronic acid
◇	Kdn	2-keto-3-deoxy-nonulosonic acid
●	Man	mannose
◣	ManN	mannosamine
◈	ManA	mannuronic acid
■	ManNAc	N-acetylmannosamine
◆	NeuAc	N-acetylneuraminic acid
◇	NeuGc	N-glycolylneuraminic acid
☆	Xyl	xylose
◹	Undef	undefined

O-glycosylated include hydroxyproline found in plants, and hydroxylysine in collagens.

Glycosphingolipids are usually attached via ceramide residues. In prokaryotes, a variety of glycoconjugates not found in mammalian systems exist, including lipopolysaccharides (LPS), which consists of three parts: a lipid A moiety embedded in the outer membrane, a core oligosaccharide containing KDO and heptose which are monosaccharides that are not found in vertebrates, and a polysaccharide side chain known as the O-antigen.

Polysaccharides forming polymers may be considered another type of class whereby repeated structures (up to millions) of monosaccharide components form very large structures. For example, cellulose fibers consisting of repeated β1-4Glc residues are coated and cross-linked with one another by glycans called hemicelluloses, of which xyloglucan is the major representative found in the primary cell walls of most higher plants. Chitin is another polymer consisting of GlcNAcβ1-4 residues and is considered the second most abundant biopolymer on Earth, next to cellulose. It is a major component of the exoskeleton of arthropods, and it is also found in the cell-wall of fungi as well as the cuticle of nematodes. Peptidoglycans are another type of bacterial polymer constituting the major structural component of the periplasm. It consists of MurNAcβ1-4GlcNAcβ1-4 repeat units, covalently cross-linked to short peptides.

Table 1.2: Generally recognized classes of glycans.

Class (subtype) name and description	Core structure
N-linked: Commonly found in mammalian systems and further subdivided into three sub-classes (below).	
N-linked (high-mannose): Synthesized early in the N-linked glycan biosynthetic pathway, then later trimmed down to form one of the other two subtypes.	

Continued on next page...

Table 1.2 – Continued

Class (subtype) name and Description	Core structure
N-linked (complex): After the high-mannose structure has been trimmed down to the tri-mannose core structure, GlcNAcs are added to both antennae, resulting in the complex subtype.	
N-linked (hybrid): This subtype contains mannoses on one antenna while the other has GlcNAcs, like the complex subtype.	
O-linked (core 1): This subtype comprises most O-linked glycans.	
O-linked (core 2): This subtype requires the O-linked core 1 structure as a substrate and is the most commonly found *in vivo*.	
O-linked (core 3): This subtype is not commonly found in many tissues, except for the intestinal tract, where mucin production is normally high. The enzyme forming this subtype, Core 3 GlcNAcT, is believed to compete with the enzyme forming the core 1 subtype, Core 1 GalT.	
O-linked (core 4): Similarly to the core 2 subtype, this subtype requires the core 3 structure as a substrate.	

Continued on next page...

Table 1.2 – Continued

Class (subtype) name and Description	Core structure
O-linked (core 5): Cores 5 through 8 are rare O-glycan subtypes that have only been found in a few selected mucins, in particular related to carcinoma.	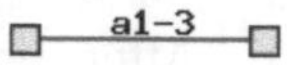 a1–3
O-linked (core 6):	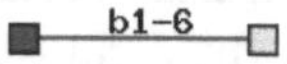 b1–6
O-linked (core 7):	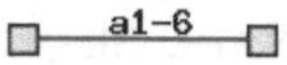 a1–6
O-linked (core 8):	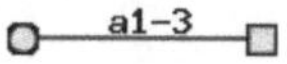 a1–3
GAG (Hyaluronic acid or hyaluronate (HA)): This is a non-sulfated repeated structure of 10^4 disaccharides on average, and it is often found in skin and skeletal tissues as well as the vitreous of the eye, umbilical cord and synovial fluid.	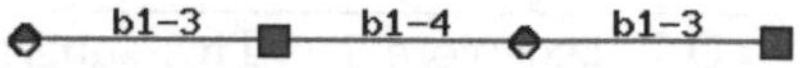 b1–3 b1–4 b1–3
GAG (Chondroitin sulfate (CS)): This is a sulfated repeated structure, linked to xylose bound to serine residues on core proteins. The sulfates are not drawn here, but the GalNAcs can be sulfated in a variety of patterns, including 2-, 4-, 2- and 6-, and 4- and 6-.	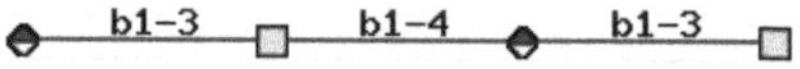 b1–3 b1–4 b1–3
GAG (Dermatan sulfate (DS)): Formerly called chondroitin sulfate B since this structure is the same as CS except for the epimerisation of glucuronic acid to iduronic acid, which takes place after CS is formed.	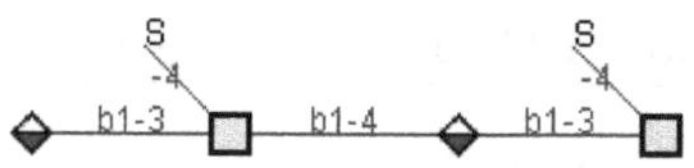 b1–3 b1-4 b1-3

Continued on next page...

Table 1.2 – Continued

Class (subtype) name and Description	Core structure
GAG (Keratan sulfate (KS)): This structure is a linear polymer of Galβ1-4GlcNAc repeats, sulfated at the C6 of both hexose moieties. It is classified further into KSI and KSII based on the linkage to the core protein. KSI is N-linked to Asn residues whereas KSII is O-linked to Ser or Thr residues.	b1-4 b1-3 b1-4 (with S / -6 sulfation)
GAG (Heparin): This structure is the same as HS, except that it has a higher degree of sulfation and is found only in mast cells.	a1-4 b1-4 a1-4 (S/-6, -3, 2-S, NS sulfation)
GAG (Heparan sulfate (HS)): This is a sulfated repeated structure, linked to xylose bound to serine residues on core proteins. It is made by virtually all cells.	a1-4 b1-4 a1-4 (S/-6, -2, NS sulfation)
GSL (arthro series): This class of GSLs are characteristically found in insects (the name deriving from Arthropoda). It is thought that this GSL series corresponds to to the ganglioside series in vertebrates. (Sugita et al. (1989))	b1-4 b1-3 b1-4
GSL (gala series): The major neutral GSL found in mollusca sea snails, but in general a less commonly found GSL since its core structure contains a Gal-Cer as opposed to a GlcCer (ie., the reducing end is a galactose as opposed to a glucose). (Itonori and Sugita (2005))	a1-4
GSL (ganglio series): This is the core structure shared by the ganglio-series GSLs. The two residues at the non-reducing end may repeat to produce longer structures. (R.K. Yu (2007))	b1-3 b1-4 b1-4

Continued on next page...

Table 1.2 – Continued

Class (subtype) name and Description	Core structure
GSL (globo series): This is a neutral GSL that has been found in the liver fluke, *Fasciola Hepatica*. The two residues at the non-reducing end of this core structure may be repeated to produce longer GSLs. The base core structure does not include the GalNAc. (Itonori and Sugita (2005); R.K. Yu (2007))	□—a1-3—○—a1-4—○—b1-4—●
GSL (isoglobo series): Similar to the globo series, the base core structure does not include the GalNAc. This structure may also extend with the two residues at the non-reducing end repeated.	□—b1-3—○—a1-3—○—b1-4—●
GSL (lacto series): The basic core structure does not include the galactose at the non-reducing end. However, the disaccharide structure at the non-reducing end can be repeated to generate longer structures.	○—b1-3—■—b1-3—○—b1-4—●
GSL (mollu series): This class of GSLs are characteristically found in Mollusca, thus the name. (Itonori and Sugita (2005))	■—b1-2—◒—a1-3—◒—b1-4—●
GSL (muco series): A less common structure of which little is known.	○—b1-3—○—b1-4—○—b1-4—●
GSL (neolacto series): Similar to the lacto series except that the final galactose residue is β1-4 linked as opposed to β1-3 linked.	○—b1-4—■—b1-3—○—b1-4—●

Continued on next page...

Table 1.2 – Continued

Class (subtype) name and Description	Core structure
LPS: Found in the outer membrane of eubacteria, LPS consists of a lipid A moiety, which is embedded in the outer membrane, and two carbohydrate components that extend outward: (i) a core oligosaccharide containing monosaccharides not found in vertebrates (such as the two rectangles KDO and the black rectangle heptose) and (2) a polysaccharide side chain known as the O-antigen. The blob located on the core is PP-Etn (pyrophosphoethanolamine).	

Abbrev: GAG = glycosaminoglycan; GSL = glycosphingolipid

1.4 Glycan biosynthesis

Glycans are synthesized by enzymes called glycosyltransferases which transfer a sugar residue from an activated nucleotide sugar donor to specific acceptor molecules, forming glycosidic bonds (Breton et al. (2006)). Transfer of the sugar residue occurs with either the retention or the inversion of the configuration of the anomeric carbon (McNaught and Wilkinson (1997)). These enzymes can be found in both prokaryotes and eukaryotes, having high specificity for both the glycosyl donor as well as the acceptor substrates. In this section, the major biosynthesis pathways of glycans will be introduced.

Glycosylation produces different types of glycans that are typically attached to proteins or lipids. Protein glycosylation includes *N*-glycans, *O*-glycans and glycosaminoglycans (GAGs). Lipid glycosylation includes glycolipids (glycosphingolipids) and glycosylphosphatidylinositol (GPI)-anchors. In mammalian systems, these glycans are constructed in an ordered manner through the workings of glycosyltransferases and also glycosidases, which are enzymes that remove specific glycosidic linkages from a glycan structure.

1.4.1 *N*-linked glycans

N-glycans constitute the most well-known class of glycans, having the largest core structure. They occur on many secreted and membrane-bound glycoproteins on the Asn residue of seqons containing the consensus sequence Asn-

X-Ser/Thr, where the X can be any amino acid except for Pro. *N*-glycans may occasionally occur at Asn-X-Cys, provided that the cysteine residue is in the reduced form. The identity of the X residue may affect the efficiency of *N*-glycosylation, such as when it is an acidic residue (aspartate or glutamate). Thus not all such sequons containing this consensus sequence may be *N*-glycosylated; they are only considered potential *N*-glycosylation sites, and experimental evidence using glycomics technologies, for example, are necessary to confirm glycosylation.

The *N*-glycan biosynthesis pathway is illustrated in Figure 1.7. In mammalian systems it begins on the cytosolic side of the endoplasmic reticulum (ER) membrane, with GlcNAc and mannose activation to build up the lipid-linked oligosaccharide precurser (LLO), which is assembled on dolichol (Dol). Dol in the ER is activated to dolichol phosphate by a kinase. A GlcNAc-1-phosphotransferase (ALG7), a GlcNAc-transferase (ALG13 or ALG14), and two mannosyltransferases ALG1 and ALG2 then consecutively build up the *N*-glycan core structure. Then an α(1-2)mannosyltransferase ALG11 adds two additional mannoses to the α(1-3) branch of the tri-mannosyl core. A flippase then flips the structure across the membrane bilayer such that it lies on the lumen side of the ER. Next, four mannoses are added by more enzymes ALG3, ALG9, ALG12 and ALG9, followed by three glucoses by ALG6, ALG8 and ALG10. This Glc3Man9GlcNAc2 structure is now ready to be transferred to asparagine residues on nascently translated proteins. The three glucoses of the transferred structure are consecutively removed by glucosidases. It appears that these glucoses are required as a quality control flag for transfer to the protein. Additionally, mannoses are removed until the structure is reduced to a Man5GlcNAc2 configuration, by which time the structure has moved to the Golgi. As an alternate route, it may happen that the Glc1Man5GlcNAc2 structure escapes early into the Golgi, whereby an endo α-mannosidase reduces the structure to Man5GlcNAc2, a hybrid-type *N*-glycan. The GlcNAc transferase I enzyme (MGAT1) then adds a GlcNAc to the α(1-3)Man on the tri-mannosyl core, which is then reduced to Man3GlcNAc3 by α-mannosidase II. This structure may also be obtained by an alternative route, where α-mannosidase II first reduces the Man5GlcNAc2 structure to Man3GlcNAc2, onto which GlcNAc transferase I adds a GlcNAc. From this Man3GlcNAc3 structure, GlcNAc transferase II may add an additional GlcNAc to the tri-mannosyl core, starting the complex *N*-glycan biosynthesis pathway. Alternatively, sugars may be attached such that the structures form hybrid type *N*-glycans. Most secreted and cell surface *N*-glycans are of the complex type (Varki et al. (1999)).

Further additions of monosaccharides occurring in the *trans*-Golgi may be divided into three components: (1) additions to the core, (2) elongation of branched terminal GlcNAc residues, and (3) capping of the elongated branches. In (1), the main core modification is the α1-6 fucosylation of the first GlcNAc on the chitobiose core, mainly in vertebrate *N*-glycans. In invertebrates, glycoproteins may have up to four fucose residues on both GlcNAcs

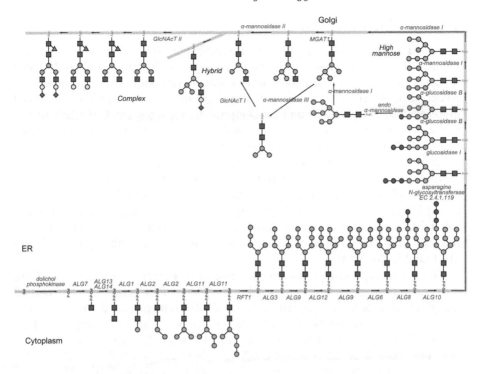

FIGURE 1.7: The *N*-glycan biosynthesis pathway.

in the core structure either α1-3 or α1-6 linkages. In plants, only the α1-3 fucosylation on the first GlcNAc occurs. Additionally, a bisecting GlcNAc may be added by GlcNAc-T III which adds a GlcNAc to the first βMan in a β1-4 linkage, preventing the action of α-mannosidase II, thus resulting in the biosynthesis of hybrid type *N*-glycans. Another common modification to the core structure is the addition of xylose in β1-2 linkage to the βMan of the core, mainly found in plants. In (2), the majority of hybrid and complex type *N*-glycans are elongated by the addition of type-2 LacNAc, which is a β-linked Gal residue to the initiating GlcNAc residue, producing the ubiquitous building block of Galβ1-4GlcNAc. Multiple additions of LacNAc produce tandem repeats of these structures. Alternatively, type-1 LacNAc structures (Galβ1-3GlcNAc) may be added, and some glycoproteins may contain LacdiNAc extensions (GalNAcβ1-4GlcNAc), but these two types of structures are not often found in tandem repeats. Finally, in (3), after (2), *N*-glycan structures are capped with sialic acids, Fuc, Gal, GalNAc or sulfate in α-linked conformations such that further elongation does not occur. This conformation facilitates the presentation of terminal monosaccharides to lectins and antibodies (Varki et al. (2008)).

1.4.2   *O*-linked glycans

In eukaryotes, *O*-linked glycans are assembled on either a serine or threonine residue of a peptide chain in the Golgi apparatus. However, unlike *N*-linked glycans, there is no known consensus sequence, although a proline residue at either -1 or +3 relative to the Ser/Thr appears to be favorable for *O*-linked glycosylation. Glycans bound in *O*-glycosidic linkages via GalNAcα to the Ser or Thr residues of proteins are called mucin-type *O*-glycans. Other types of *O*-linked glycans include Manα-Ser/Thr in yeast and mammalian proteins, GlcNAcβ-Ser/Thr in nuclear and cytoplasmic proteins, GlcNAcα attached to hydroxyproline in cytosolic proteins from *Dictyostelium*, and Glcβ-Ser/Thr and Fucα-Ser/Thr in blood clotting factors. Xylβ-Ser linked *O*-glycans are found in proteoglycans (heavily glycosylated glycoproteins having a core protein with one or more attached glycosaminoglycan (GAG) chains) and are thus introduced in the section on glycosaminoglycans. Galβ-hydroxy-Lys linkages are found in collagens.

The mucin-type *O*-glycans are classified into several core structures. Core 1 is characterized by Galβ1-3GalNAc1-Ser/Thr. This structure is extended by an addition of a GlcNAcβ1-6 to the GalNAc to form core 2. Alternatively, core 3 is the structure GlcNAcβ1-3GalNAcα1-Ser/Thr, becoming core 4 by an addition of GlcNAcβ1-6 to the GalNAc. Core 5 is GalNAcα1-3GalNAcα1-Ser/Thr, core 6 is GlcNAcβ1-6GalNAcα1-Ser/Thr, core 7 is GalNAcα1-6 GalNAcα1-Ser/Thr, and core 8 is Galα1-3GalNAcα1-Ser/Thr. These core structures may be further substituted by fucose and/or sialic acid; they may also be elongated by repeated Galβ1-4GlcNAc or Galβ1-3GlcNAc groups, similar to *N*-glycans (Kobata (2007)).

O-mannose biosynthesis begin with the activity of the POMT1-POMT2 mannosyltransferase complex, which adds αmannose to Ser/Thr. POMGNT1 then adds a β1-2GlcNAc residue to the mannose. The *O*-mannose structure is completed by the addition of a Galβ1-4, followed by NeuAcα2-3. Other branches may also be generated with the addition of glucuronic acid, sulphate, etc.

1.4.3   Glycosaminoglycans (GAGs)

Glycosaminoglycans, or GAGs, are mucopolysaccharides; they are long unbranched polysaccharides consisting of a repeating disaccharide unit. These sugars may be sulfated in various positions, and the degree of sulfation is believed to affect function. The major GAGs are keratan sulfate, chondroitin sulfate, dermatan sulfate, heparan sulfate, heparin and hyaluronan.

GAGs contain a core structure of GlcAβ1-3Galβ1-3Galβ1-4Xylβ1-Ser, from which chondroitin sulfate/dermatan sulfate (CS) and heparan sulfates/heparin (HS) can be generated. CS is produced by the addition of the disaccharides GlcAβ1-3GalNAcβ1-4 which may be sulfated. GalNAcT-I transfers the GalNAcβ1-4 to the non-reducing end of the chain, while GlcAT-II transfers

TABLE 1.3: Core structures of glycosphingolipids. Key differences are underlined.

Glycosphingolipid series	Core structure
ganglio-series	Galβ1-3GalNAcβ1-4Galβ1-4GlcβCer
lacto-series	Galβ1-3GlcNAcβ1-3Galβ1-4GlcβCer
neolacto-series	Galβ1-4GlcNAcβ1-3Galβ1-4GlcβCer
globo-series	Galα1-4Galβ1-4GlcβCer
isoglobo-series	Galα1-3Galβ1-4GlcβCer

the glucuronic acid. The GalNAc may be 4-0 or 6-0 sulfated by corresponding sulfotransferases. Some of the GlcA residues may also be epimerized to IdoA, which may be 2-0 sulfated. HS, on the other hand, is elongated by GlcAβ1-4GlcNAcα1-4 by EXTL.2 and EXT1/EXT2. EXTL.2 adds a single GlcNAcα1-4 residue to the core tetrasaccharide, and EXT1 and EXT2 alternatingly add GlcAβ1-4 and GlcNAcα1-4 residues. A GlcNAc *N*-deactylase/*N*-sulfotransferase then acts on a portion of the GlcNAc residues in a cluster along the chain to add sulfates. Most of the *N*-deacetylated GlcNAc residues are concomitantly *N*-sulfated. An epimerase then acts on a portion of the GlcA residues adjacent to GlcNS, forming IdoA, which are 2-0 sulfated. Some of the remaining GlcA residues are also 2-0 sulfated. The GlcNS and GlcNAc residues may also be 3-0 and 6-0 sulfated.

1.4.4 Glycosphingolipids (GSLs)

Glycosphingolipids, or GSLs, are known to be involved in defining antigenic specificities of some cells. They have been shown to function as (a) cell type-specific and developmental stage-specific antigens and (b) isogenic or heterophile antigens, such as histo-blood group antigens (Hakomori and Igarashi (1995)).

GSLs are attached to ceramide residues, forming either a glucosylceramide (GlcCer) or galactosylceramide (GalCer) core (R.K. Yu (2007)). Most mammalian GSLs including sialic acid-containing GSLs called gangliosides contain the former, while the relatively small gala-series contains the latter. GlcCer GSLs are divided into three major classes: the ganglio-series, the lacto- and neolacto-series, and the globo- and isoglobo-series. The core structures of these glycosphingolipids are listed in Table 1.3.

GSLs are further subclassified into three subclasses: neutral (having no charged sugars or ionic groups), sialylated (having sialic acid residues), or sulfated. Sialylated GSLs are traditionally called gangliosides, regardless of their core structure. Other common, but unofficial, names for GSLs include GM1, for example, which refers to the ganglioside series (G), the number of sialic acid residues (M=mono, D=di, T=tri, etc.), and the order of migration of the ganglioside on thin-layer chromatography. Thus GM3 > GM2 > GM1.

As for invertebrates, insects and mollusks, for example, have GSLs having completely different core structures: Manβ1-4GlcβCer, which is called the *arthro* core structure, and the major sphingolipids in fungi and plants are inositolphosphate ceramides.

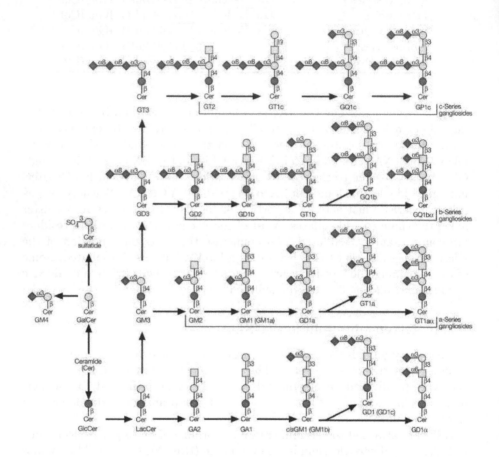

FIGURE 1.8: The biosynthesis pathway of glycosphingolipids in the brain. *Reused by permission of Consortium of Glycobiology Editors, La Jolla, California.*

The biosynthesis pathway of brain GSLs is illustrated in Figure 1.8 (Varki et al. (2008)). First, either UDP-Gal:ceramide β-galactosyltransferase or UDP-Glc:ceramide β-glucosyltransferase transfers either a Gal or Glc, respectively, to ceramide. GalCer sulfotransferase then adds a sulfate group to the C-3 of galactose on GalCer to form sulfatide. In contrast, UDP-Gal:GlcCer β1-4 galactosyltransferase transfers a Gal to GlcCer to form lactosylceramide, or LacCer, which is followed by CMP-NeuAc:lactosylceramide α2-3 sialyltransferase which forms the ganglioside GM3. From GM3, different glycosyltransferases compete with one another to form various structures. UDP-GalNAc:GM3/GD3 β1-4 *N*-acetylgalactosaminyltransferase may generate a-series gangliosides, whereas CMP-NeuAc:GM3 α2-8 sialyltransferase generates GD3 and the b-series gangliosides. Similarly, an α2-8 sialyltransferase generates GT3 from GD3 to form the c-series gangliosides.

For the initial stages of GSL biosynthesis, GSL-specific enzymes are involved. However, in later stages, glycosylntransferases that act on glycoproteins also act on GSLs such that similar terminal structures are formed. For example, the α1-3 *N*-acetylgalactosaminyltransferase encoded by the blood group A gene and the α1-3 galactosyltransferase encoded by the allelic blood group B gene act on both glycoproteins as well as glycolipids.

1.4.5 GPI anchors

Proteins attached to glycosyl-phosphatidylinositol (GPI) anchors via their carboxyl termini are normally found on the outer leaflet of the lipid bilayer facing the extracellular environment. This GPI-anchored form of proteins allow them to be rendered soluble compared to transmembrane proteins. In some eukaryotic microbes such as the yeast *S. cerevisiae*, GPI anchors are used to target certain mannoproteins for covalent incorporation into the β-glucan cell wall. Without GPI anchors, defects would occur in cell-wall biosynthesis which are known to be detrimental to yeast. GPI biosynthesis is also essential for the bloodstream form of *T. brucei*, where nutritional stress occurs without the essential GPI-anchored transferrin receptor (Varki et al. (2008)).

The core structure of GPI anchors is constructed by the addition of glucosamine to the C-6 of the inositol ring of phosphatidylinositol. Manβ1-4, Manα1-6, and Manα1-2 are then sequentially attached to the glucosamine residue to form a linear structure. A phosphoethanolamine residue is then attached to the C-6 position of the last mannose residue, and this phosphoethanolamine is then linked to the amino acid at the C-terminal of a protein.

1.4.6 LPS

Lipopolysaccharides (LPS) are recognized by the innate immune system, stimulating inflammatory responses in order to clear bacteria that have breeched the barrier defenses, such as the skin and mucosal epithelium. LPS biosynthesis starts with the formation of Lipid A: First, UDP-GlcNAc is acylated at C3, followed by N-deacetylation, N-acylation, and cleavage of the pyrophosphate linkage, forming 2,3-diacylglucosamine-1-P, as in Figure 1.9. Two of these molecules condense, forming the tetra-acyl disaccharide core, which is then phosphylated on C4' of the nonreducing sugar and modified with Kdo. This assembly process occurs in the inner membrane, and the resulting structure is flipped across the membrane by a transporter to face the periplasm. A number of transferases then further add heptoses, glucoses and phosphate groups to form the core region. The O-antigen is assembled independently of the core structure, and the entire antigen structure is transferred at once to the core region of LPS. The entire complex is then transferred to the outer membrane (Varki et al. (2008)).

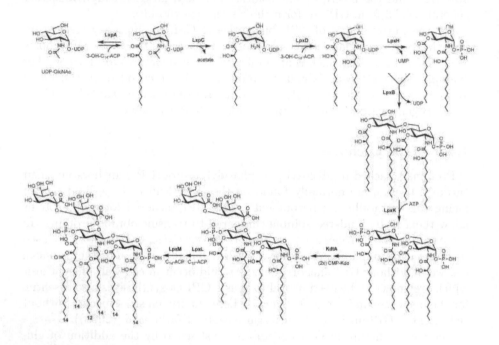

FIGURE 1.9:　The biosynthesis pathway of LPS. *Reused by permission of Consortium of Glycobiology Editors, La Jolla, California.*

1.5　Glycan motifs

Glycan classes are organized according to the core structure in general. However, there are also common structural patterns that are often found on the terminal end of glycans that may be found across glycan classes. These motifs are described in this section (Varki et al. (2008)).

The structures often found at the terminal ends of N-glycans were briefly introduced earlier. One of the most common terminal structures is Galβ1-4GlcNAc, which is known as **LacNAc**, for N-acetyllactosamine. This disaccharide may repeat to form a chain of LAcNAcs, termed poly-N-acetyllactosamine, which are found in glycans from most cell types. The **LacdiNAc** structure is an alternative to LAcNAc, composed of GalNAcβ1-4GlcNAc units. These lactosamine units are also called type-2 glycan units. Type-1 glycan units, on the other hand, are disaccharides called neo-*N*-acetyllactosamine, which are Galβ1-3GlcNAc structures. Little is as of yet known about this structure.

Poly-*N*-acetyllactosamine chains correspond to the "i" blood group antigen. These chains may be branched by the addition of GlcNAc in β1-6 linkage to the internal Gal residues, forming chains corresponding to the "I" blood group antigen. These structures are illustrated in Figure 1.10.

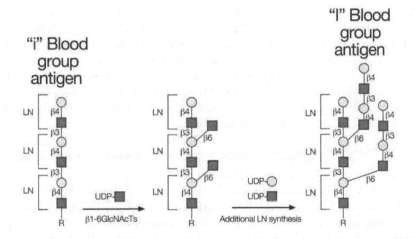

FIGURE 1.10: The "I" and "i" blood group antigens, consisting of poly-*N*-acetyllactosamine chains. *Reused by permission of Consortium of Glycobiology Editors, La Jolla, California.*

The ABO blood group antigens in human are formed by the tissue-specific glycosylation of these branched and unbranched poly-*N*-acetyl-lactosamine chains in either type-1 or type-2 form. The addition of fucose in α1-2 linkage to the Gal in type-1 or type-2 LacNAc forms the **blood group H determinant**. Fucose attached to type-2 and type-4 (glycolipid) units form the **H antigen** on red cells. Fucose transferred to type-1 and type-3 (O-GalNAc) LacNAcs form the **H antigen** in epithelia.

After transfer of fucose, the A or B blood group determinants can be formed. The **blood group A** glycan epitope is formed by the addition of GalNAc in an α1-3 linkage to the terminal Gal. The **blood group B** glycan determinant is formed by the transfer of Gal in α1-3 linkage instead. Thus these glycan structures determine the blood type of individuals, with blood group

A individuals synthesizing the A determinants, blood group B individuals synthesizing the B determinants, blood group AB individuals synthesizing both, and blood group O individuals synthesizing neither.

The Lewis blood group antigens are a related set of glycans carrying either α1-3 or α1-4 fucose residues attached to the GlcNAc of type-1 neo-*N*-acetyllactosamine. This basic structure is called the Lewis^a antigen (Le^a). The terminal Gal on Le^a may be modified by fucose in α1-2 linkage, forming the Lewis^b antigen (Le^b). These structures can be further sialylated in a α2-3 linkage or sulfated at the C3 position of the terminal Gal, forming sialyl Le^a or 3'-sulfo-Le^a, respectively. Other members of the Lewis blood group family include Lewis^x (Le^x) and Lewis^y (Le^y) determinants. Le^x is formed by the transfer of the fucose to the GlcNAc of the type-2 LacNAc structure, and Le^y is formed by the further addition of fucose in α1-2 linkage to the terminal Gal. Sialyl Le^x and 3'-sulfo-Le^x may also be formed by the sialylation in a α2-3 linkage or the sulfation of the C3 position of the terminal Gal of Le^x. The sialyl-Le^x structures may also be further sulfated at either the C6 position of the GlcNAc or the Gal, forming 6-sulfo-sialyl Lewis^x (6-sulfo-SLe^x) or 6'-sulfo-sialyl Lewis^x (6'-sulfo-SLe^x). Both of these may also be sulfated simultaneously to form 6,6'-bisSulfo-sialyl Lewis^x (6,6'-bisSulfo-SLe^x). Figure 1.11 illustrates these Lewis structures.

Glycolipids of the red cell membrane have P blood group antigens expressed on them. The **P^k antigen** is constructed by the addition of Gal in an α1-4 linkage to **lactosylceramide** (Galβ1-4GlcβCer). An additional transfer of GalNAcβ1-3 to the terminal Gal of the P^k antigen forms the **P antigen**. Alternatively, a GlcNAc may be transferred to the terminal Gal on lactosylceramide in a β1-3 linkage to form lactotriaosylceramide, which can form paragloboside by the addition of a Gal in a β1-4 linkage. Paragloboside is the substrate for the **P1 antigen**, which contains an additional Gal in an α1-4 linkage.

1.6 Potential for drug discovery

The biosynthesis process of glycans was described in Section 1.4. If defective, many of the key enzymes in this process are known to cause disorders called congenital disorders of glycosylation (CDG). During the first five years of life, only about 20% of patients survive. However, the mortality rate decreases as the patient ages. The CDGs currently known may be caused at almost all steps of the *N*-glycan biosynthesis pathway by the inactivation of the dolichol phosphokinase, ALG1, ALG2, ALG3, ALG9, ALG12, ALG6, ALG8, glucosidase I, among many others. The key features of CDGs include mental retardation and epilepsy.

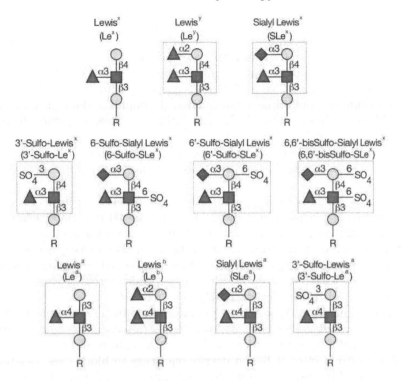

FIGURE 1.11: The Lewis blood group antigens. *Reused by permission of Consortium of Glycobiology Editors, La Jolla, California.*

O-glycosylation is initiated with the addition of GalNAc to Ser/Thr by GALNT3, whose mutations have been found to be linked to familial tumoral calcinosis, a severe autosomal recessive metabolic disorder involving massive calcium deposits in the skin and subcutaneous tissues. T-synthase is a galactosyltransferase that adds Galβ1-3 to the GalNAc core. T-synthase requires an X-linked gene called COSMC for proper folding and normal activity, and mutations in COSMC are known to cause the rare autoimmune disease Tn syndrome.

In the *O*-mannose glycosylation pathway, mutations in POMT1 and POMT2 have been found in many of the patients with Walker-Warburg syndrome, an extremely severe form of congenital muscular dystrophy. Furthermore, mutations in POMGNT1 cause muscle-eye-brain disease, characterized by symptoms similar to Walker-Warburg syndrome.

There several human disease phenotypes known that are related to genes involved in the biosynthesis of GAGs. For example, the extra-cellular matrix defects that cause bone and cartilage abnormalities in the progeriod variant of Ehlers-Danlos syndrome are known to be a result of mutations in B4GALT7, which adds the first Galβ1-4 to the xylose in the core. Hereditary multiple exostosis is characterized by bony outgrowths (exostoses), and is known to be caused mostly by mutations in EXT1 and EXT2.

In glycosphingolipids, mutations in the gene encoding ST3GAL5 have been identified as the cause of Amish infantile epilepsy syndrome. This is caused by the accumulation of multiple non-sialylated glycolipids in these patients. During GPI-anchor biosynthesis, impaired synthesis of these structures is known to cause paroxysmal nocturnal haemoglobinuria (PNH), in which the progeny of abnormal multipotent haematopoietic stem cells lacking GPI-anchored proteins become resistant to apoptosis, thus dominating the population (Freeze (2006)).

Not only is there potential for therapy targeting inherent genes in the system, the role of glycans in recognition by bacterial pathogens, for example, is also important, and there are currently glycoconjugate vaccines against three pathogens currently available. Many others are also seeking licenses. The licensed vaccines work against Hib, *N. meningitidis* and multiple pneumococcal serotypes. The first were the Hib vaccines, which virtually eliminated Hib meningitis in the 1980s (Jones (2007)).

LPS contains lipid A, which is known as endotoxin, a potent stimulator of innate immunity. This molecule contributes to secondary complications of infections, including septic shock, multiple organ failure, and mortality. Thus a major research interest lies in developing drugs to block these deleterious effects exerted by this compound during pathogenesis. Many aminoglycoside antibiotics such as penicillins and cephalosporins are inhibitors of peptidoglycan biosynthesis (Varki et al. (2008)).

Glycosylation also plays a key role in virus infection. For example, influenza viruses cause infection through the glycosylation patterns on both the host cell receptor and the two main viral membrane proteins. Neuraminidase, which cleaves the sialic acid from membrane glycolipids, assists virus particles to be released from host cells. Thus this neuraminidase has been the target of influenza drug therapies.

Glycans are also potential biomarkers for cancer. Many tumor antigens are actually glycoproteins or glycolipids, whose monoclonal antibodies were generated against peptide portions of the glycoprotein or sugar portions of the glycolipid. For example, the sialic Lewis structures are cancer biomarkers for a wide variety of cancers (Kannagi (2004); Itai et al. (1988)). Thus many clinical cancer diagnostic tests use these glycoproteins as cancer markers. These are however not specific to a particular cancer on the whole. A few tissue-specific markers do exist, however, such as alpha-fetoprotein for primary hepatoma and prostate specific antigen for prostate cancer (Packer et al. (2008)).

Chapter 2

Background

This chapter provides background information regarding glycome informatics, which will be required for understanding the methods introduced in this book. First, the nomenclature for representing glycans in computerized form will be described. There are numerous formats that are already being actively used by various groups. Although there is currently work on standardizing these formats, knowledge about these formats are essential in order to use the respective databases from which these formats developed. The latter part of this chapter focuses on glycan-related interactions, including an introduction to lectins and their interactions, and carbohydrate-carbohydrate interactions.

2.1 Glycan nomenclature

Due to the development of glycan databases at around the same time but independently of one another, several formats for representing glycan structures have been developed. Many of these formats represent glycans using a connection table, or adjacency matrix. These data structures allow the definition of glycans as graphs, where a list of nodes (representing monosaccharides) is given, and the list of connections, or edges (representing glycosidic bonds), describe the glycan structure. The major formats for representing glycans are described in this section.

2.1.1 InChI[TM]

The IUPAC International Chemical Identifier (InChI[TM], pronounced "IN-chee") is a non-proprietary identifier for chemical structures, proposed as a standard for encoding such structures in databases (Stein et al. (2003)). InChI describes chemical structures as hierarchical layers of information, including the atoms and their bonds, isomer information, isotope information, stereochemistry, and electronic charge information. Only those layers that are applicable to the structure need be specified. Layers and sublayers are both separated by a slash (/) character, and each sublayer is listed after its preceding (parent) layer. Except for the chemical formula sublayer, all layers and sublayers start with a prefix character. The five layers and important sublayers are as follows:

1. Main layer (M)

   - Chemical formula (no prefix): The only required sublayer; the conventional Hill sorted elemental formula.[1] If multiple components exist, the Hill-sorted formulas of each component are sorted and separated by dots.
   - Atom connections (prefix c): The bonds between the atoms in the structure, partitioned into at most three sublayers concerning hydrogen atoms (H-atoms) (prefix h):
     (a) All bonds other than those to non-bridging H-atoms
     (b) Bonds of all immobile H-atoms
     (c) Locations of any mobile H-atoms, representing those that may be found at more than one location in a compound due to isomerization.

2. Charge layer

   - The net charges of the components (prefix q)
   - The number of protons removed from or added to the substance such that a given component may be represented regardless of its degree of protonation (prefix p)

3. Stereochemical layer

   - Double bond sp^2 (Z/E) stereo (prefix b)
   - Tetrahedral sp^3 stereo: relative sp^3 stereochemistry is represented first, optionally followed by a tag to indicate absolute stereochemistry. If unknown, an "unknown" descriptor may be specified. An "undefined" flag may also be given if no stereo information is provided but a stereocenter may be possible. Three possible prefixes may be used: t for sp^3, m for inverted sp^3, and s for the type (1=abs, 2=rel, 3=rac).

4. Isotopic layer (MI): different isotopically labeled atoms; isotopic hydrogen atoms that are interchangeable (such as deuterium and tritium) are listed separately. Any changes in stereochemistry caused by the presence of isotopes are also listed here. Prefixes include i for isotopic atoms, h for exchangeable hydrogen atoms, and the same four prefixes that

---

[1]The Hill System Order defines the order by which to specify a chemical compound. The number of carbon atoms are listed first, the number of hydrogen atoms second, and all other elements in alphabetical order. If no carbon is present, then all elements are listed in alphabetical order. The number is indicated immediately after each element symbol.

are used in the stereochemical layer, referring to the stereochemistry of isotopic atoms.

5. Fixed H layer (F): optional; potentially mobile H atoms which should be immobile are specified here. Any changes to earlier layers due to this specification are also added to this layer.

A sixth Fixed/Isotopic Combination (FI) layer may also be optionally added at the end, referring to isotopic fixed H atoms. Prefixes include i for isotopic fixed H atoms, o for transpositions, and the four prefixes used for the stereochemical layer.

There are also a number of auxiliary prefixes that may accompany the InChI code. Details are provided in the InChI Technical manual which is included with the software, downloadable from: http://www.iupac.org/inchi/. As an example, the InChI format for glucose, illustrated in Figure 1.3 is written as 1/C6H12O6/c7-1-2-3(8)4(9)5(10)6(11)12-2/h2-11H,1H2.

2.1.2 (Extended) IUPAC format

Although the two-dimensional notation of glycans as in Figure 1.2 may be visually appealing, it is not suitable for storage in a database, and bioinformatic analysis tools would not be able to make use of it. Thus the IUPAC-IUBMB (International Union of Pure and Applied Chemistry - International Union of Biochemistry and Molecular Biology) has specified the "Nomenclature of Carbohydrates" to uniquely describe complex oligosaccharides based on a three-letter code to represent monosaccharides. For example, *gal* represents galactose and *man* represents mannose; a listing of the common monosaccharides (and their derivatives) that occur in oligo- and polysaccharides is given in Table 2.1 (Tsai (2007)). Each monosaccharide code is preceded by the anomeric descriptor and the configuration symbol. The ring size is indicated by an italic *f* for furanose or *p* for pyranose. The carbon numbers that link the two monosaccharide units are given in parentheses between the symbols separated by an arrow. For example, the structure in Figure 1.2 would be represented as: α-D-Man*p*-(1→3)[α-D-Man*p*-(1→6)]-α-D-Man*p*-(1→4)- β-D-Glc*p*NAc-(1→4)-β-D-Glc*p*NAc. Double-headed arrows may be used if monosaccharides are linked through their anomeric centers. Moreover, α and β may be represented as a or b, respectively. In such a way, long carbohydrate sequences can be adequately described in abbreviated form using a sequence of letters. This format is called the extended IUPAC form.

Table 2.1: Common monosaccharides (and their derivatives) that occur in oligo- and polysaccharides.

Monosaccharide	IUPAC format	Representative occurrence
D-Ribose	D-Rib*f*	β-D-Rib*f* is the sugar component of RNA.
2-Deoxy-D-Ribose	D-dRib*f*	D-dRib*f* is the sugar component of DNA.
D-Xylose	D-Xyl*f*	Xylan, found in the hemicellulose of plants, consists of this monosaccharide in β1-4 linkages with branches in β1-3. D-Xyl*f* is attached to threonine in glycoproteins.
L-Arabinose	L-Ara*f*	Arabinan consists of this monosaccharide in α1-5 linkages with side chains in α1-3. L-Ara*f* is found with GalA in pectin and with Xyl in plant cell walls.
D-Glucose	D-Glc*p*	D-Glc*p* is the most abundant monosaccharide, constituting many oligo- and polysaccharides, forming sucrose, maltose and lactose. It also makes up cellulose, starch, glycogen and dextran, in a variety of linkages.
D-Galactose	D-Gal*p*	D-Gal*p* makes up lactose and galactans.
D-Mannose	D-Man*p*	This monosaccharide makes up the *N*-glycan core structure in β1-4, α1-3 and α1-6 linkages. It can also be found in α1-2 linkages further along the chain. In *O*-glycoproteins, it is attached to serine residues.
D-Fructose	D-Fru*f*	Often found in polymers, β-D-Fru*f*-(1$\leftrightarrow$2)-α-D-Glc*p* forms sucrose. This monosaccharide also makes up fructan in α2$\rightarrow$1 and α2-6 linkages.

Continued on next page...

Table 2.1 – Continued

Monosaccharide	IUPAC format	Representative occurrence
L-Fucose	L-Fuc*p*	L-Fuc*p* may also be denoted as 6-deoxy-L-Gal. This monosaccharide makes up blood group polysaccharides as well as glycoconjugates via α1-4 and α1-6 linkages.
L-Rhamnose	L-Rha*p*	This monosaccharide makes up the *O*-antigen structure of LPS, and is also found in gums and mucilages.
D-Glucuronic acid	D-Glc*p*A	A derivative of glucose, this monosaccharide unit makes up chondroitin and hyaluronic acid together with D-GlcNAc. GlcA is also found in hemicellulose, gums and mucilages.
D-Galacturonic acid	D-Gal*p*A	This monosaccharide forms the polymer pectin in α1-4 linkages.
N-acetyl-D-glucosamine	D-Glc*p*NAc	GlcNAc is linked to asparagine in *N*-glycans and may be found in β1-2, β1-4 and β1-6 linkages. This monosaccharide makes up many GAGs such as chondroitin sulfate, heparin and hyaluronic acid.
N-acetyl-D-galactosamine	D-Gal*p*NAc	D-GalNAc is found in mucopolysaccharides such as chondroitin, keratan sulfates and dermatan sulfates. This monosaccharide is linked to serine in *O*-glycoproteins.
N-acetylmuramic acid	D-Mur*p*Ac	This monosaccharide may also be denoted as 2-acetamido-2-deoxy-3-*O*-[(R)-1-carboxyethyl]-D-glucose. This monosaccharide makes up the peptidoglycan of the bacterial cell wall together with GlcNAc in β1-4 linkages.

Continued on next page...

Table 2.1 – Continued

Monosaccharide	IUPAC format	Representative occurrence
N-acetylneuraminic acid	D-Neu*p*Ac	This monosaccharide is called sialic acid and may also be denoted as 5-acetamido-3,5-dideoxy-D-glycero-D-galacto non-2-ulosonic acid. It is usually linked to the terminal ends of glycans in α2-3 or α2-6 linkages.
N-glycolylneuraminic acid	D-Neu*p*Gc	This monosaccharide is also called sialic acid, but it is found mostly in non-human mammalian species such as horse, sheep, and ape. It is believed that this sugar was lost during human evolution for reasons that are as of yet unclear.

2.1.3   CarbBank format

IUPAC also suggests an extended IUPAC form by which structures are written across multiple lines. This is the format originally used by CarbBank, thus it is sometimes referred to as such. The representation of monosaccharides is the same as that of IUPAC format, where each monosaccharide residue is preceded by the anomeric descriptor and the configuration symbol and the ring size is indicated by an italic *f* or *p*. If any of α/β, D/L or *f*/*p* are omitted, it is assumed that this structural detail is unknown.

As an example, the *N*-glycan core structure in Figure 1.2 would be represented as in Figure 2.1. This format may substitute α and β with a and b,

α –D–Man*p*– (1→6) +

↓

β –D–Man*p*– (1→4) – β –D–Glc*p*NAc– (1→4) – α –D–Glc*p*NAc

↑

α –D–Man*p*– (1→3) +

FIGURE 2.1:   The *N*-glycan core structure represented in CarbBank (extended IUPAC) format.

respectively. Arrows ($\rightarrow$) may also be replaced by hyphens (-), and up ($\uparrow$) and down ($\downarrow$) arrows may be replaced by bars (|).

2.1.4 KCF format

The KEGG Chemical Function (KCF) format for representing glycan structures was originally used to represent chemical structures (thus the name) in KEGG. KCF uses the graph notation, where nodes are monosaccharides and edges are glycosidic linkages. Thus to represent a glycan, at least three sections are required: ENTRY, NODE and EDGE. An example of the *N*-glycan core structure in KCF format is displayed in Figure 2.2.

```
ENTRY       XYZ         Glycan
NODE        5
            1       GlcNAc    15.0        7.0
            2       GlcNAc     8.0        7.0
            3       Man        1.0        7.0
            4       Man       -6.0       12.0
            5       Man       -6.0        2.0
EDGE        4
            1       2:b1       1:4
            2       3:b1       2:4
            3       5:a1       3:3
            4       4:a1       3:6
///
```

FIGURE 2.2: The *N*-glycan core structure represented in KCF format.

The ENTRY line in Figure 2.2 may optionally contain a name for the glycan (in this case, XYZ). The NODE keyword is followed by the number of residues represented in the structure. Then the following lines list the monosaccharide names, one line on each row, numbered from one (1). Each of these rows are prefixed by ten spaces and specify monosaccharides in the following format: "*n name x y*" where *n* is the *n*th residue in the list, *name* is the freely specifiable residue name, and *x* and *y* specify the *x* and *y* coordinates by which to draw the glycan structure in 2D space.

The EDGE keyword follows the NODE section, first specifying by the number of edges (normally one less than the number of nodes for tree stuctures). Similar to the NODE section, the EDGE keyword is followed by rows prefixed by spaces defining the glycosidic linkages in the structure in the following format: "*e* n_1:ac_1 n_2:c_2" where *e* is the *e*th glycosidic linkage in the list, n_1 and

TABLE 2.2:   Keywords used in the KCF format, listed in the order in which they are normally displayed.

Keyword	Format	Explanation
ENTRY	*name* `Glycan`	Name of the glycan (optional)
COMPOSITION	(*res1*)*num1* (*res2*)*num2* ...	Monosaccharide residue composition
MASS	*float* (*aglycons*)	Mass of monosaccharides and additional aglycons
CLASS	*class*;*subclass*	Hierarchical classification of the given glycan
REACTION	*Rnum1 Rnum2* ...	KEGG REACTION entries involving the given glycan
ENZYME	*ECnum1 ECnum2* ...	Enzyme Commission (EC) numbers involving the given glycan
DBLINKS	*dbname: dbID1 dbID2* ...	Other database entries (usually CCSD for CarbBank) of this glycan
NODE	see text	Monosaccharide and other aglycon constituents of this glycan
EDGE	see text	Linkage information contained in this glycan
BRACKET	see text	Repeating unit specifier

n_2 correspond to the monosaccharide numbers that are linked, a represents the anomeric configuration (either a or b), and c_1 and c_2 represent the carbon numbers of monosaccharides n_1 and n_2, respectively, which are linked. Finally, three slashes (///) indicate the end of the given entry.

Repeating units in KCF are specified in the BRACKET section, of which an example is given in Figure 2.3 with its corresponding structure in Figure 2.4. The BRACKET section contains the (x, y) coordinates of the four corners in the 2D plane which covers the repeating substructure. For a particular repeating subunit, three rows are specified. The first two rows use the following format: $nx_1y_1x_2y_2$ followed by $nx_3y_3x_4y_4$ where n is the nth repeating unit in the structure and x_1y_1 correspond to the top-left corner of the bracket by which to enclose the repeating subunit. x_2, y_2, x_3, y_3, and x_4, y_4 correspond to the bottom-left, bottom-right and top-right corners of the bracket, respectively. The third row of this section is specified as nm where m indicates the number of times that the subunit repeats (n if unknown).

Optional keywords that may be found in the KCF files in the KEGG GLYCAN database include MASS, COMPOSITION, and REACTION, among others. All the keywords that may appear in a KCF entry are listed and explained in Table 2.2.

2.1.5   LINUCS format

The LInear Notation for Unique description of Carbohydrate Sequences (LINUCS) format was developed such that a single ASCII string could uniquely

```
ENTRY        G11158                      Glycan
NAME         Sulfoadhesin
COMPOSITION  (Gal)3 (GalNAc)1 (GlcNAc)2 (LFuc)2 (Neu5Ac)2 (S)2 (Ser/Thr)1
MASS         2130.9 (Ser/Thr)
REFERENCE    1   [PMID:12855678]
             de Graffenried CL, Bertozzi CR.
             Golgi localization of carbohydrate sulfotransferases is a
             determinant of L-selectin ligand biosynthesis.
             J. Biol. Chem. 278 (2003) 40282-95.
NODE         13
             1    Ser/Thr     25      0
             2    GalNAc      15      0
             3    Gal          5     -8
             4    GlcNAc       3      9
             5    GlcNAc      -5     -8
             6    S           -7     14
             7    Gal         -7      9
             8    LFuc        -7      4
             9    S          -15     -3
             10   Gal        -15     -8
             11   LFuc       -15    -13
             12   Neu5Ac     -17      9
             13   Neu5Ac     -25     -8
EDGE         12
             1     2:a1     1
             2     3:b1     2:3
             3     4:b1     2:6
             4     5:b1     3:3
             5     6        4:6
             6     7:b1     4:4
             7     8:a1     4:3
             8     9        5:6
             9    10:b1     5:4
             10   11:a1     5:3
             11   12:a2     7:3
             12   13:a2    10:3
BRACKET      1 -11.0   15.0  -11.0    3.0
             1   7.0    3.0    7.0   15.0
             1 n
             2 -19.0   -2.0  -19.0  -14.0
             2   0.0  -14.0    0.0   -2.0
             2 n
///
```

FIGURE 2.3:   An example of a glycan structure containing repeating units
in KCF format.

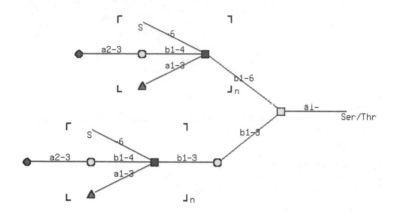

FIGURE 2.4:   The glycan structure of the KCF in Figure 2.3, containing repeating units.

define a carbohydrate structure using simple rules (Bohne-Lang et al. (2001)). It is based on the extended IUPAC format but uses additional rules to define the priority of the branches. In this way, carbohydrate structures can be defined uniquely while still containing all the information required to describe the structure.

The start of the LINUCS format may include two square brackets [], followed by the root residue name in square brackets. If a residue has a single child, then the child's linkage in parentheses surrounded by square brackets precedes the child's residue name and configuration (as in IUPAC format) in square brackets. If a residue has more than one child, then each child's branch is surrounded by curly brackets {}. Children are listed in order of the carbon number linking them to the parent, such that the child with a 1-3 linkage would come before a child with a 1-4 linkage. As an example, given the glycan structure in CarbBank format in Figure 2.5, the corresponding structure in LINUCS format is Figure 2.6. Note that the α1-3Man is listed before α1-6Man in both branches (denoted by underlined text). Structures in LINUCS format may thus be specified linearly as [] [b-D-GlcpNAc]{[(4+1)] [b-D-GlcpNAc] {[(4+1)] [b-D-Manp]{[(3+1)] [a-D-Manp]{[(2+1)] [a-D-Manp]{[(2+1)] [a -D-Manp]{}}}[(6+1)] [a-D-Manp]{[(3+1)] [a-D-Manp]{[(2+1)] [a-D-Manp]{}}[(6+1)] [a-D-Manp]{[(2+1)] [a-D-Manp]{}}}}}}.

2.1.6   BCSDB format

The BCSDB format is used in the BCSDB database (see Section 3.1.4) to encode carbohydrates and derivative structures in a single line. Residues are described in the format $<res>(<c_1>-<c_2>)$ where *res* is the name of the residue and its configuration and c_1 and c_2 correspond to the carbon numbers

```
a-D-Manp-(1-2)-a-D-Manp-(1-6)+
                              |
                 a-D-Manp-(1-6)+
                 |            |
   a-D-Manp-(1-2)-a-D-Manp-(1-3)+    b-D-Manp-(1-4)-b-D-GlcpNAc-(1-4)-b-D-GlcpNAc
                                                  |
a-D-Manp-(1-2)-a-D-Manp-(1-2)-a-D-Manp-(1-3)+
```

FIGURE 2.5: A glycan structure in extended IUPAC (CarbBank) format. Its corresponding LINUCS notation is illustrated in Figure 2.6.

```
[] [Asn] {
  [ (4+1) ] [b-D-GlcpNAc] {
    [ (4+1) ] [b-D-GlcpNAc] {
      [ (4+1) ] [b-D-Manp] {
        [ (3+1) ] [a-D-Manp] {
          [ (2+1) ] [a-D-Manp] {
            [ (2+1) ] [a-D-Manp] { }
          }
        }
        [ (6+1) ] [a-D-Manp] {
          [ (3+1) ] [a-D-Manp] {
            [ (2+1) ] [a-D-Manp] { }
          }
          [ (6+1) ] [a-D-Manp] {
            [ (2+1) ] [a-D-Manp] { }
          }
        }
      }
    }
  }
}
```

FIGURE 2.6:   The glycan structure in Figure 2.5 in LINUCS format.

of the child and parent, respectively, by which the residue *res* is linked to its parent. Of course the portion in parentheses is omitted for the residue at the root. If c_1 or c_2 are unknown, a question mark (?) may be used. If the glycan structure is a repeated unit, then parts of the portions in parentheses may be hanging at the ends, such as in -2)A(1-3)B(1-4)C(1-, which represents the repeated structure linked by a 1-2 linkage. For branched structures, it is assumed that there is only one main chain, and the rest are branches. Side chains are enclosed in square brackets together with their linkage in parentheses, as in t)A(1-3)[B(1-4)]C, indicating that residue C has two children, A and B, linked by 1-3 and 1-4 linkages, respectively. The t) prefix indicates that the residue to its right is at the non-reducing end. For more than two children, the side branches are enumerated and separated by commas within the square brackets. Side branches may also contain branched structures.

Thus the structure -4)A(1-3)[D(2-6)B(1-4),F(1-3)[G(1-4)]E(1-2)]C(1- corresponds to the structure in Figure 2.7.

```
G-(1-4)+
       |
F-(1-3)-E-(1-2)+
               |
      -4)-A-(1-3)-C-(1-
                     |
D-(2-6)-B-(1-4)+
```

FIGURE 2.7:   The structure corresponding to the glycan in BCSDB format: -4)A(1-3)[D(2-6)B(1-4),F(1-3)[G(1-4)]E(1-2)]C(1-.

Since not all residues encoded in BCSDB format are necessarily monosaccharides, it may be possible for a residue to substitute more than one position of its parent residue. In this case, the substituent should be indicated twice, separated by a comma, as in ...(1-2)[xRPyr(2-4):xRPyr(2-6]aDGal(1-..., which indicates a 4,6-pyruvated galactose residue. In the case that such a residue is at the non-reducing end, the notation becomes xRPyr(2-4)[:xRPyr (2-6]aDGal(1-.... There are also cases when the glycan structure may not be uniquely identified: (1) when a residue has exactly one of two possible children and (2) when a residue may have up to two children from a set of candidates. In case (1), the notation D(1-2)<<A(1-3)|B(1-4)>>C indicates that the structure is either of the form D(1-2)A(1-3)C or D(1-2)B(1-4)C. In case (2), <A(1-3)|B(1-4)>C indicates that it may be one of the structures from case (1) or even a trimer where parent C has both children A and B attached. Monovalent substitutions of monosaccharides, such as acetylation, are described as separate residues. Thus, aDGal(1-3)bDGlcNAc should be specified as aDGal(1-3)[Ac(1-2)]bDGlcN. At the reducing end, a monovalent residue may also be specified as aDGlc(1-Me.

As in the LINUCS format, there are several rules used in the BCSDB format in order to specify carbohydrate structures uniquely. For polymers, the polymer backbone is always considered the main chain. For oligomers or substructures, priorities are given to carbohydrate residues over monovalent substituents. In the case that both chains of a branch are monosaccharides or both are monovalent substituents, then the child with the smaller carbon number is chosen as the main chain. In the case of multiple side chains, residues are listed in descending order of carbon number, as in Ac(1-6),Ac(1-2)]. Residue names are composed of the following fields in order with no separators.

1. Anomeric configuration: a for α, b for β, 1 for a lipid residue, x for a residue with no or more than one anomer, ? for unknown configuration. Monovalent residues do not require this field.

2. Stereoisomers: One of D, L, R, S, X, or ?, where X is used when no such configuration exists and ? is used for unknown configurations. Monovalent residues do not require this field.

3. Residue base name: This includes deoxygenation. The first letter should be capitalized, and the rest in lowercase.

4. Ring size: One of p for pyranose, f for furanose, a for open-chain, or ? for unknown or any. The question mark may be omitted except when the residue name ends with one of these symbols (i.e., Ala or Rha).

5. Capital A if a residue is a uronic acid.

6. Amino group modifiers: one or more capital Ns. The position of the amino group should be specified unless it is 2. For example, aLRHA4N has a modification at position 4. If the amino group is implied by the residue base name, then this modification should not be applied.

7. -ol modifier for alditol residues (when not implied by the residue base name).

A table listing the monomer namespace is provided at the following URL: http://www.glyco.ac.ru/bcsdb/residues.php. Moreover, if a specific residue cannot be identified, a superclass name may be used instead of the residue name. For example, PEN = pentose, HEX = hexose, HEP = heptose, OCT = octose and NON = nonose.

BCSDB also contains glycolipid structures, which are beyond the scope of this text. Interested readers may refer to the following URL which describes the BCSDB format in detail: http://www.glyco.ac.ru/bcsdb/help/rules.html.

2.1.7 Linear Code®

Linear Code® is a carbohydrate format defined by GlycoMinds, Ltd. It uses a single-letter nomenclature for monosaccharides and includes a condensed description of the glycosidic linkages. Monosaccharide representation is based on the common structure of a monosaccharide (listed in Table 2.3), where modifications to the common structure are indicated by specific symbols, as in the following (Banin et al. (2002)):

- Stereoisomers (D or L) differing from the common isomer are indicated by apostrophe (').

TABLE 2.3:   List of common monosaccharide structures and their single-letter code as used in the Linear Code® format. Note that all the sugars are assumed to be in pyranose form unless otherwise specified.

Common configuration	Full name	Linear Code®
D-Glc*p*	D-Glucose	G
D-Gal*p*	D-Galactose	A
D-Glc*p*NAc	*N*-Acetylglucosamine	GN
D-Gal*p*NAc	*N*-Acetylgalactosamine	AN
D-Man*p*	D-Mannose	M
D-Neu*p*5Ac	*N*-Acetylneuraminic acid	NN
D-Neu*p*	Neuraminic acid	N
KDN[2]	2-Keto-3-deoxynanonic acid	K
Kdo	3-deoxy-D-manno-2 Octulopyranosylono	W
D-Gal*p*A	D-Galacturonic acid	L
D-Ido*p*	D-Ioduronic acid	I
L-Rha*p*	L-Rhamnose	H
L-Fuc*p*	L-Fucose	F
D-Xyl*p*	D-Xylose	X
D-Rib*p*	D-Ribose	B
L-Ara*f*	L-Arabinofuranose	R
D-Glc*p*A	D-Glucuronic acid	U
D-All*p*	D-Allose	O
D-Api*p*	D-Apiose	P
D-Fru*f*	D-Fructofuranose	E

[2] 3-deoxy-D-glycero-K-galacto-nonulosonic acid

TABLE 2.4:   List of common modifications as used in the Linear Code® format.

Modification Type	Linear Code®
deacetylated *N*-acetyl	Q
ethanolaminephosphate	PE
inositol	IN
methyl	ME
N-acetyl	N
O-acetyl	T
phosphate	P
phosphocholine	PC
pyruvate	PYR
sulfate	S
sulfide	SH
2-aminoethylphosphonic acid	EP

- Monosaccharides with differing ring size (furanose or pyranose) from the common form are indicated by a caret (^).

- Monosaccharides differing in both of the above are indicated by a tilde (˜).

For example, D-Gal*p* is the common form of this monosaccharide, so its code A is used alone. To specify L-Gal*p* instead, A' is used. For D-Gal*f*, A^ and for L-Gal*f*, A˜.

Modifications to the residues listed above are represented by adding square brackets including the connecting position of the modification followed by the modification symbol (listed in Table 2.4) without any separators. Thus D-Gal*p* with sulfate in the third position would be written as A[3S]. Multiple modifications are written in numerical order by position within the same brackets. The only exceptions to this rule apply to common modifications that are listed in Table 2.3 such as *N*-acetylgalactosamine, which can be represented as A[2N] but is instead represented as simply AN.

Linkage information is represented using the symbols a and b for α and β, respectively. This is followed by the carbon number of the parent to which the residue is attached. Thus the structure β-D-Gal*p*(2P)-(1-3)-β-D-Glc*p* would be written as A[2P]b3Gb. When a modification occurs at the first carbon, the number 1 is usually omitted. Furthermore, if the sugar at the reducing end is in its open form (ol), then the letter o is added. To handle branches, parentheses "()" are used, similar to how curly brackets were used for the LINUCS format. Repeating and cyclic units can also be handled by Linear Code®. A cyclic motif is represented by the letter c, and repeating units are written inside curly brackets as {n}, where n represents the number of repeats. For example, cellulose, a polymer of repeated β(1-4)Glc residues, would be written as {nGb4}. If the repeating units are not connected at both ends, then the monosaccharide at which the unit is connected is marked surrounded by dashes, as in {nGa6Ga4(-Ab3-)Ub2Ha3Ha3Ha3}.

Linear Code® also allows the specification of glycoconjugates by using different symbols depending on the type of conjugate. Amino acid sequences are written after a semicolon (;). For example, α-D-Glc bound to Asn-Tyr-Ser-Cys would be written as Ga;NYSC. If necessary, the amino acid to which the glycan is attached may be indicated by surrounding dashes, as in Ga;NY-S-C. Lipid moieties are indicated after a colon (:), using the Linear Code® representation for lipids: C: ceramide, D: sphingosine, IPC: inositolphosphoceramide, and DAG: diacylglycerol. Third, other types of glycosides are written after the number (or pound) symbol (#) using its complete name, as Gnb3Ab#4-Trifluoroacetamidophenol.

Unknown or uncertain information can also be handled by Linear Code®. For example, if a specific detail of a linkage is unknown, the question mark (?) can be used, as in AN?3G for an unknown anomer. If the connection position is also unknown, then the 3 above would be replaced by a question mark as well. If even the residue is unknown, then asterisk (*) is used, as in ANb3*A

TABLE 2.5:   List of monosaccharides and their
three-letter codes used in GlycoCT.

Monosaccharide name	Three-letter code	Superclass
Allose	ALL	HEX
Altrose	ALT	HEX
Arabinose	ARA	PEN
Erythrose	ERY	TET
Galactose	GAL	HEX
Glucose	GLC	HEX
Glyceraldehyde	GRO	TRI
Gulose	GUL	HEX
Idose	IDO	HEX
Lyxose	LYX	PEN
Mannose	MAN	HEX
Ribose	RIB	PEN
Talose	TAL	HEX
Threose	TRE	TET
Xylose	XYL	PEN

which indicates a trimer whose middle residue is unknown. If two possibilities
are given for a linkage, a slash (/) is used to separate the two candidates, as
in ANb3/4, which indicates that the linkage may be β1-3 or β1-4. For two
possible residues, a double slash (//) is used to separate the two candidates,
as in Ab4//Ga2Aa3 to indicate one of the following two structures: Ab4Aa3 or
Ga2Aa3. Oftentimes the residue(s) at the non-reducing end of a structure may
be linked to two or more possible antennae. In this case, a variable of two
characters is used, a percentage symbol preceded by a number. Thus 1% and
2% indicate two separate uncertainties. The uncertain residues are written
at the end of the Linear Code® following a vertical bar (|). For example,
NNa6=1%|1%Ab4GNb2Ma3(1%Ab4GNb2Ma6)Mb4Gb indicates that the sialic acid
residue may be linked to the non-reducing ends of either antennae at the
positions indicated by 1%.

2.1.8   GlycoCT format

GlycoCT was developed as a part of the EuroCarbDB project, and it is used
as the main format in the GlycomeDB database (Section 3.1.7). GlycoCT uses
a similar graph concept to the KCF format (Herget et al. (2008)) and consists
of two varieties: a condensed format and an XML format. The former allows
for unique identification of glycan structures in a compact manner, while the
latter facilitates data exchange. While these will be described in general here,
more detailed explanations can be found in the online GlycoCT handbook at
http://www.eurocarbdb.org/recommendations/encoding.

The monosaccharide namespace consists of five components and basically
follows those defined by IUPAC: the basetype, anomeric configuration, the

TABLE 2.6:   List of substituents used in GlycoCT.

acetyl	amidino	amino
anhydro	bromo	chloro
diphospho	epoxy	ethanolamine
ethyl	fluoro	formyl
glycolyl	hydroxymethyl	imino
iodo	lactone	methyl
N-acetyl	N-alanine	N-amidino
N-dimethyl	N-formyl	N-glycolyl
N-methyl	N-methyl-carbomoyl	N-succinate
N-sulfate	N-triflouroacetyl	nitrate
phosphate	phospho-choline	phospho-ethanolamine
pyrophosphate	pyruvate	succinate
sulfate	thio	triphosphate
(r)-1-hydroxyethyl	(r)-carboxyethyl	(r)-carboxymethyl
(r)-lactate	(r)-pyruvate	(s)-1-hydroxyethyl
(s)-carboxyethyl	(s)-carboxymethyl	(s)-lactate
(s)-pyruvate	(x)-lactate	(x)-pyruvate

monosaccharide name with configurational prefix, chain length indicator, ring forming positions and further modification designators. Trivial names such as fucose or rhamnose are not permitted in GlycoCT. The monosaccharide naming convention follows the following format: $a-bccc-DDD-e : f|g : h$, where a is the anomeric configuration (one of a, b, o, x), b is the stereoisomer configuration (one of d, l, x), ccc is the three-letter code for the monosaccharide as listed in Table 2.5, DDD is the basetype or superclass indicating the number of consecutive carbon atoms such as HEX, PEN, NON, e and f indicate the carbon numbers involved in closing the ring, g is the position of the modifier, and h is the type of modifier (one of d=deoxygenation, a=acidic function, keto=carbonyl function, en= double bond, aldi=reduction of C1-carbonyl, sp2=outgoing double bond linkage, sp=outgoing triple bond linkage, geminal=two identical substitutions). For a, b, e, f and g, an x can be used to specify an unknown value. $bccc$ and $g : h$ may also be repeated if necessary. Thus α-D-Gal$p$ would be represented as a-dgal-HEX-1:5 and α-D-Kdn$p$ would be a-dgro-dgal-non-2:6,1a,2:keto,3:d in GlycoCT format.

It is noted that substituents of monosaccharides are also treated as separate residues attached to the base residue. These substituents are distinguished by specifying one of the following codes immediately after the residue number: b=basetype, s=substituent, r=repeating unit, a=alternative unit. The list of substituents handled by GlycoCT is given in Table 2.6.

The glycosidic linkages in GlycoCT are modeled as atom replacements, formatted as the following: $L : np(a + b)mc$, where L indicates the Lth linkage in the structure, n and m are the residue numbers of the parent and child, respectively, p and c are the parent and child atom replacement

RES
1b:a-dgal-HEX-1:5
2s:n-acetyl
3b:b-dgal-HEX-1:5
4r:r1
5b:a-dgro-dgal-NON-2:6|1:a|2:keto|3:d
6s:n-acetyl
7r:r2
8b:a-dgro-dgal-NON-2:6|1:a|2:keto|3:d
9s:n-acetyl
LIN
1:1d(2+1)2n
2:1o(3+1)3d
3:3o(3+1)4n
4:4n(3+2)5d
5:5d(5+1)6n
6:1o(6+1)7n
7:7n(3+2)8d
8:8d(5+1)9n
REP
REP1:13o(3+1)10d=-1--1
RES
10b:b-dglc-HEX-1:5
11s:n-acetyl
12b:a-lgal-HEX-1:5|6:d
13b:b-dgal-HEX-1:5
14s:sulfate
LIN
9:10d(2+1)11n
10:10o(3+1)12d
11:10o(4+1)13d
12:10o(6+1)14n
REP2:18o(3+1)15d=-1--1
RES
15b:b-dglc-HEX-1:5
16s:n-acetyl
17b:a-lgal-HEX-1:5|6:d
18b:b-dgal-HEX-1:5
19s:sulfate
LIN
13:15d(2+1)16n
14:15o(3+1)17d
15:15o(4+1)18d
16:15o(6+1)19n

FIGURE 2.8: The glycan in Figure 2.4, containing repeating units, in GlycoCT format.

identifiers, respectively, and a and b are the parent and child attachment positions, respectively. Thus for residue number 1 preserving its oxygen atom linked via its O-6 to the C-1 of residue 2 (deoxygenated), the linkage (say the first) would be specified as 1:1o(6+1)2d.

The GlycoCT format follows something similar to the KCF format, where the residues are specified in a RES section, and the linkages in a LIN section. Other sections in the GlycoCT format specify repeating units, undetermined structures, alternative structures, and non-carbohydrate entities as REP, UND, ALT, and NON, respectively. This last NON section is actually not implemented in the GlycomeDB database, but it is allowed for backward compatibility. A numerical identifier in the RES section labeled with rx indicates that the given residue position corresponds to repeating unit numbered x in the REP section. The REP section itself contains a RES and LIN section to describe the repeating substructure, numbered sequentially following the original RES numbering order. The first line of the REP section is specified as follows: REPx:$c(l)p$=n-m, where x is the xth repeating unit, c and p are the numbers of the residues toward the non-reducing end and reducing ends of the substructure being repeated, l is the linkage information, and n-m indicate the range of numbers by which to repeat. Thus the glycan structure containing repeats in Figure 2.4 can be specified in GlycoCT format as in Figure 2.8.

Alternative units which describe two potential substructures in the middle of a glycan are represented using the ALT section. Within the ALT section, multiple ALT subgraphs may be specified, so subsections numbered ALT1, ALT2, ... are defined. Within a single ALTn section, the potential substructures are defined within ALTSUBGRAPHm subsections, and within a single ALTSUBGRAPHm subsection, residues and linkages are specified using RES and LIN, respectively. Here, a line in the format LEAD-IN RES:x and an optional line in the format LEAD-OUT RES:y should also be specified to indicate the reducing end and optional non-reducing end residues by which to attach to the main structure. As was for the REP section, the numbering of the residues in the RES sections should follow after the numbering of residues in the main RES and REP sections.

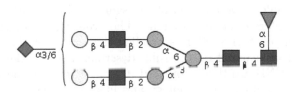

FIGURE 2.9: A glycan structure with an undetermined residue.

Undetermined terminal units are handled in the UND section, to describe residues or structures known to exist but whose linkages are undetermined. There are also undetermined residues in the middle of the structure, such as sulfation patterns which indicate sulfation by a certain percentage of probability. The UND section thus contains a line to indicate the probability of occurrence of the undefined structure. An example is given in Figure 2.9 where the terminal sialic acid is known to exist, but whose linkage is undetermined. The corresponding GlycoCT would be described as in Figure 2.10.

Similar to the LINUCS rules for specifying unique carbohydrate sequences, GlycoCT also incorporates some hierarchical rules by which the ordering of residues, linkages, and special features is unambiguously defined. First, the main RES section appears first, followed by the LIN section (which may be omitted if no linkages are defined). Subsequent sections appear in the following order: REP, UND, ALT and NON. Because the GlycoCT format allows for the definition of structures of possibly unconnected trees, rules for prioritizing the structures are determined by the following order: (a) the number of child residues, (b) the length of the longest branch, (c) the number of terminal residues, (d) the number of branching points, and (e) the lexical order. Thus the ordering of all trees in a GlycoCT representation can be uniquely defined.

To order branches, the following rules are checked in order: (a) the number of bonds between parent and child residues, (b) the atom linkage position of the parent, (c) the atom linkage position of the child, (d) the linkage type of the parent, (e) the linkage type of the child, and (f) the result of comparing the child residues using the residue comparison rules listed above. Here, the linkage type can be any of the following values:

b basetype

s substituent

n non-carbohydrate unit

r repeating unit

a alternative unit

In this way, unless two children of the same residue are both unknown or identical, all children can be ordered. The ordering of the RES and LIN sections within the ALT section is handled similarly. For the UND section, each UND is sorted based on the rules described above. The reducing end residues of all UND are then compared using the residue rules. In the case where they are identical, the topology and linkages of the parent residues of the UND residues are evaluated by (a) comparing the list of parent residues from each UND and (b) comparing the parent linkages (the linkage between the UND and the main graph) from each UND.

```
RES
1b:x-dglc-HEX-1:5
2s:n-acetyl
3b:b-dglc-HEX-1:5
4s:n-acetyl
5b:b-dman-HEX-1:5
6b:a-dman-HEX-1:5
7b:b-dglc-HEX-1:5
8s:n-acetyl
9b:b-dgal-HEX-1:5
10b:a-dman-HEX-1:5
11b:b-dglc-HEX-1:5
12s:n-acetyl
13b:b-dgal-HEX-1:5
14b:a-lgal-HEX-1:5|6:d
LIN
1:1d(2+1)2n
2:1o(4+1)3d
3:3d(2+1)4n
4:3o(4+1)5d
5:5o(3+1)6d
6:6o(2+1)7d
7:7d(2+1)8n
8:7o(4+1)9d
9:5o(6+1)10d
10:10o(2+1)11d
11:11d(2+1)12n
12:11o(4+1)13d
13:1o(6+1)14d
UND
UND1:100.0:100.0
ParentIDs:9|13
SubtreeLinkageID1:o(3|6+1)d
RES
15b:a-dgro-dgal-NON-2:6|1:a|2:keto|3:d
16s:n-acetyl
LIN
14:15d(5+1)16n
```

FIGURE 2.10: The glycan in Figure 2.9, containing an undetermined component, in GlycoCT format.

2.1.9   XML representations

Several XML representations for characterizing glycan structures to ease data transfer and exchange have been developed. The earliest representation is the CabosML format (Kikuchi et al. (2005)), which was followed by GlycoCT and GLYDE (Sahoo et al. (2005)). CabosML and GLYDE are based on abstracted monosaccharide residues, representing carbohydrate structures in a tree structure formalism as opposed to a connection table or adjacency list. This limits the types of structures that can be represented in these formats, considering that cyclic glycan structures are known to exist. The encapsulation of the conjugates with which the glycans are bound is also important. Thus, the GLYDE-II XML format was developed and proposed as the standard XML data exchange format for carbohydrate structures (Packer et al. (2008)) and will be discussed here.

The syntax of GLYDE-II is defined in the document type definition (DTD) of the standard, which can be found at the following URL: http://glycomics. ccrc.uga.edu/GLYDE-II/. The ELEMENT types defined by GLYDE-II are listed in Table 2.7; the root ELEMENT being the GlydeII type, which may consist of one or more of the following ELEMENT types: free_atom, molecule, or aggregate. An aggregate may contain a mixture of different ELEMENT types, referencing previously defined free_atoms and/or other aggregates. Combining these types, they are referred to as free_parts since none of them are linked to one another. An aggregate may also contain a model_ref, which is a reference to an externally defined ELEMENT and may only be used at the top level.

A molecule is composed of a set of uniform ELEMENTs consisting of bound_atoms, residues or moieties and corresponding links atom_link, res-idue_link or moiety_link, respectively. Note that a molecule consists of the same types of ELEMENTs linked to their corresponding link types. Thus molecules containing bound_atom ELEMENTs may only contain bound_atom ELEMENTs linked only by atom_link ELEMENT types. The molecule EL-EMENT contains the following attributes: id, subtype (one of glycoprotein, glycopeptide, glycolipid, peptidoglycan, glycan, protein, peptide, li-pid, monosaccharide, or amino_acid), a name, a boolean radical value (yes/ no, default being no, used for molecules with unpaired electrons), and a value for the charge state (for ions, which have non-zero charge).

A free_atom is an atom that is not linked to any other structure; it may be a model_ref. Its attributes are id, name, a boolean radical value (yes/no, default being no), and a value for the charge state.

A moiety consists of residues, but it is defined as a reference to a molecule which serves as its archetype. Thus its attributes consist of partid, subtype (one of glycan, protein, peptide or lipid), ref and name. In contrast, a residue consists of bound atoms, and it is also a reference to another archetype molecule. In general, a residue may represent a monosaccharide (called base_type), a substituent, an amino acid or a lipid. Thus its at-

TABLE 2.7: The ELEMENT types defined in GLYDE-II.

ELEMENT	Description
GlydeII	Root ELEMENT.
model_ref	A reference to an external source defining the ELEMENT.
moiety	A reference to a molecule serving as its archetype, which can be one of glycan, protein, peptide or lipid.
residue	A molecule component representing a monosaccharide, a substituent, an amino acid or a lipid.
bound_atom	A molecule component that is not a moiety or residue.
moiety_link	Linkage ELEMENT between two moieties.
residue_link	Linkage ELEMENT between two residues.
atom_link	Linkage ELEMENT connecting two bound_atoms, which includes not only the connection information, but also the atoms that were replaced in order to form the link.
molecule	A set of uniform ELEMENTs consisting of bound_atoms, residues or moieties and their corresponding links atom_link, residue_link or moiety_link.
aggregate	A mixture of different unlinked ELEMENTs.
free_atom	An atom that is not linked to any other structure; possibly a model_ref.
free_part	An instantiation of a previously defined archetype, which can be a molecule, free_atom, or aggregate. It composes an aggregate.
combination	A molecule consisting of combinations of more than one of moiety_link, residue_link or atom_link, thus enabling the representation of possible links between parts that are defined in a molecule containing the combination.
repeat_block	An ELEMENT specifying the residues, moieties, bound_atoms or nested repeat_blocks that are repeated in tandem along with the repeated links in between.
repeat_residue	A component of a repeat_block.

tributes consist of partid, subtype (one of lipid, amino_acid, base_type or substituent), ref and name. The third molecule subtype is bound_atom, which is linked to other bound_atoms in a molecule. It is defined as a reference to a free_atom which serves as its archetype. Its attributes consist of partid, ref, name, parity (referring to its stereochemistry, either -1/1), and InChIatom, which specifies the InChI numbering of the atom within the context of the molecule.

A free_part ELEMENT composes an aggregate. It is not connected to any other ELEMENT and is an instantiation of a previously defined archetype (molecule, free_atom, or aggregate). Its attributes consist of partid, type (one of free_atom, molecule, or aggregate), subtype (one of glycoprotein, glycopeptide, glycolipid, peptidoglycan, glycan, protein, peptide, lipid, monosaccharide, or amino_acid), and ref, which corresponds to the id of its archetype.

The ELEMENTs moiety_link and residue_link connects two moietys or two residues, respectively. They each contain the following attributes: from, to, and stat, which specifies the probability that the link exists in the molecule. On the other hand, the ELEMENT atom_link connects two bound_atoms, which includes not only the connection information, but also the atoms that were replaced in order to form the link. Thus its attributes include from, to, from_replaces, to_replaces, bond_order (referring to the number of bonds between the two atoms), and parity, which is specified using the InChI bond parity notation to, say, differentiate cis and trans double bond structures.

A combination ELEMENT may consist of combinations of more than one of the following: moiety_link, residue_link or atom_link. A repeat_block ELEMENT specifies the residues, moieties, bound_atoms or nested repeat_blocks that are repeated in tandem along with the repeated links in between. It contains attributes repeat_number_min and repeat_number_max, which correspond to the minimum and maximum number of times the repeat_block is repeated, respectively. The repeat_number_max attribute is optional. A repeat_residue is a component of a repeat_block, and it contains a single attribute ref which is the partid of the residue that is repeated. As an example, the *N*-glycan core structure represented in GLYDE-II format is given in Figure 2.11.

2.2   Lectin-glycan interactions

Lectins are carbohydrate-binding proteins of nonimmune origin. The functions of lectins range from cell adhesion, cell recruitment, intracellular trafficking to immune recognition, and their carbohydrate recognition mechanisms

```
<GlydeII>
     <molecule subtype="glycan" id="M3N2" name="pentaglycoside">
       <residue subtype="substituent" partid="1"
              ref="http://www.monosaccharideDB.org/GLYDE-II.jsp?G=n-acetyl"/>
       <residue subtype="substituent" partid="2"
              ref="http://www.monosaccharideDB.org/GLYDE-II.jsp?G=n-acetyl"/>
       <residue subtype="base_type" partid="3"
              ref="http://www.monosaccharideDB.org/GLYDE-II.jsp?G=b-dglc-HEX-1:5"/>
       <residue subtype="base_type" partid="4"
              ref="http://www.monosaccharideDB.org/GLYDE-II.jsp?G=b-dglc-HEX-1:5"/>
       <residue subtype="base_type" partid="5"
              ref="http://www.monosaccharideDB.org/GLYDE-II.jsp?G=b-dman-HEX-1:5"/>
       <residue subtype="base_type" partid="6"
              ref="http://www.monosaccharideDB.org/GLYDE-II.jsp?G=b-dman-HEX-1:5"/>
       <residue subtype="base_type" partid="7"
              ref="http://www.monosaccharideDB.org/GLYDE-II.jsp?G=b-dman-HEX-1:5"/>
       <residue_link from="1" to="3">
         <atom_link from="N1" to="C2" from_replaces="O2" bond_order="1"/>
       </residue_link>
       <residue_link from="2" to="4">
         <atom_link from="N1" to="C2" from_replaces="O2" bond_order="1"/>
       </residue_link>
       <residue_link from="4" to="3">
         <atom_link from="C1" to="O4" to_replaces="O1" bond_order="1"/>
       </residue_link>
       <residue_link from="5" to="4">
         <atom_link from="C1" to="O4" to_replaces="O1" bond_order="1"/>
       </residue_link>
       <residue_link from="6" to="5">
         <atom_link from="C1" to="O3" to_replaces="O1" bond_order="1"/>
       </residue_link>
       <residue_link from="7" to="6">
         <atom_link from="C1" to="O6" to_replaces="O1" bond_order="1"/>
       </residue_link>
     </molecule>
</GlydeII>
```

FIGURE 2.11: The *N*-glycan core structure represented in GLYDE-II format.

are crucial for their biological functions. Some lectins are known to recognize carbohydrates monovalently; that is, a single glycan structure is recognized and bound by a single lectin. In contrast, other lectins are known to bind carbohydrates multivalently. Thus the understanding of these mechanisms is important for clarifying their roles in biological systems.

2.2.1 Families and types of lectins

Lectins have been found in almost all classes and families of organisms. Some major lectins are classified according to the glycan structures that they are known to recognize, as listed in Table 2.8. Animal lectins are characterized by a carbohydrate recognition domain (CRD), first identified in selectins (Drickamer (1988)). At least 12 structural families of animal lectins have been published (Kilpatrick (2002)), the major ones including the C-type lectin superfamily, galectins and siglecs. However, not all animal lectins may necessarily be categorized into these families. Other non-animal lectins such as those found in plants and bacteria are usually classified according to their organism taxonomy. Most plant lectins recognize a single monosaccharide and are thus useful as tools to analyze the glycan structures in unknown biological samples. Fungal lectins are also specific usually for mono- and di-saccharides.

Although many lectins that recognize monosaccharides have been studied, findings have shown that this seemingly simple interaction is more complex than it appears. It has been found that monosaccharides that are structurally dissimilar but have similar topological features may be recognized by the same lectin, such as wheat germ agglutinin (WGA), which recognizes GalNAc, GlcNAc, and Neu5Ac. However, it has also been shown that conserved sequences in homologous lectins have different specificities, and moreover that structurally different lectins may bind to identical glycan structures through different sets of residues. Thus, although the term *lectin* is used on the whole to represent carbohydrate- binding proteins, it must be noted that it is not possible to explain their interactions with glycans in general and that specificities will need to be explained at the level of individual lectin families (Sharon and Lis (2007)).

Annexins are a family of calcium- and phospholipid-binding proteins, of which over 20 members have been found in a wide variety of organisms. Their structural features include diversity in the N-terminal domains whereas the C-terminal regions consist of four or eight fairly conserved α-helical domains of approximately 70 amino acids. It is this region that binds to their ligands (Mollenhauer (1997)). The carbohydrate-binding annexins include annexin IV, V and VI, which have been shown to bind sialoglycoproteins and glycosaminoglycans in the presence of calcium. It is suggested that annexin IV is involved in the formation of apical secretory vesicles due to interactions with GPI-anchored glycoproteins and proteoglycans (Kojima et al. (1996)).

Another major family of animal lectins are the C-type lectins, which are calcium dependent and function based on the highly conserved CRD of this

TABLE 2.8: Lectin Families

Lectin Family: subfamily	Glycan ligands
Annexin IV, V, VI	Calcium- and phospholipid-binding family; annexins IV, V, VI show carbohydrate-binding activity
C-type lectin: asialoglycoprotein and DC receptors	Mainly Gal, but also Fuc and Man, with calcium
C-type lectin: collectins	Various, with calcium
C-type lectin: lecticans	Fuc, Gal, sulfated glycolipids, GlcNAc, with calcium
C-type lectin: selectins	Various, with calcium
C-type lectin: type II transmembrane receptors	Various, with calcium
Chi-lectins (Chitinase-like lectins)	Chito-oligosaccharides
F-box lectins	GlcNAc2
F-type lectins (fucolectins)	Fuc-terminating oligosaccharides
Ficolins, mannose-binding lectins	GlcNAc, GalNAc
Galectins	β-Galactosides
I-type lectins (siglecs)	Sialic acid
Intelectins (X-lectins)	Gal, galactofuranose, pentoses, with calcium
L-type lectins	Various
M-type lectins	Man8
P-type lectins	Man 6-phosphate, others
R-type lectins	Various
Tachylectins	Various (LPS, D-GlcNAc, D-GalNAc)
Leczymes: sialic acid-binding lectins (SBLs)	Sialoglycoproteins and gangliosides

family. It is known that although C-type lectins require calcium for binding, calcium itself is not directly involved, but rather provides stability for lectin function. C-type lectins may be subdivided into two types: soluble C-type lectins including lecticans and collectins, and transmembrane C-type lectins which include selectins and type II receptors. The macrophage mannose receptor, Langerin and DC-SIGN are members of the asialoglycoprotein and DC receptors subgroup. The mannose receptor is a type I transmembrane protein with an extracellular domain containing eight different C-type CRDs. It binds Man, GlcNAc and Fuc, which are not common in terminal positions of mammalian glycans, but are often found on the surfaces of microorganisms. It is suggested that the receptor thus plays a role in facilitating antigen uptake and processing in the adaptive immune response, as well as mediating direct uptake of pathogens in the innate immune response (Weis et al. (1998)). DC-

SIGN is responsible for HIV particle transfer and infection of T-cells. Langerin is an endocytic receptor and binds mannose-group monosaccharides (Zelensky and Greedy (2005)). Collectins are a subfamily of soluble lectins that participate in host-defense, having an amino-terminal collagen domain and a carboxyl terminal C-type lectin domain. Mannose-binding lectins (MBLs), or mannan-binding lectins, are members of this subfamily and play key roles in innate immunity. Lecticans are another subfamily of soluble lectins, consisting of some proteoglycan core peptides such as versican, aggrecan, neurocan and brevican, containing a single C-type lectin domain near their carboxyl termini. Selectins are a subfamily of the transmembrane C-type lectins, which function as cell-cell adhesion molecules between leukocytes and vascular endothelial cells, required for leukocyte extravasation. Selectins on leukocytes are called L-selectins, while those on the endothelial cells are called E- and P-selectins. Type II receptors are another subfamily of the transmembrane lectins, having a single transmembrane domain with a CRD located at the extracellular carboxyl terminus. Natural killer cell receptors and CD23 are typical examples of this type.

The chitinase-like family of lectins includes the protein YKL-40, which binds to chito-oligosaccharides. It was found that YKL-40 specifically binds to type 1 collagen and modulates the rate of type I collagen fibril formation, and that it in fact has no chitinase activity (Bigg et al. (2006)). These glycoproteins are found in both vertebrates and invertebrates, and they are structurally related to the family 18 glycohydrolases (as defined in CAZy, described in Section 3.2.4) which cleave the GlcNAcβ1-4GlcNAc linkage of the chitin core (Ling and Recklies (2004)).

F-box lectins are F-box proteins that target glycoproteins in a lectin-like manner (Yoshida et al. (2002)). It has been proposed that they target misfolded *N*-glycoproteins for degradation by the proteosome. They are a part of what is called the F-box associated (FBA) family (Glenn et al. (2008)), a small family of the F-box proteins. Two of the five proteins comprising the FBA family, FBX02 and FBX06, are predicted to bind high-mannose substrates through their conserved carboxyl-terminal domain, called the FBA or G domain, which is homologous to the CRDs of galectins and PNGase F (Mizushima et al. (2004)). Among the other three proteins in this family, FBX017 was suggested to bind to sulfated glycans because of its preference for heparin. It was also shown to bind lactoferrin, but no binding was detected for CS or high-mannose glycans.

F-type lectins, or fucolectins, are another small family of lectins specific for fucose, with members found in both prokaryotes as well as eukaryotes, including vertebrates and invertebrates (Honda et al. (2000)). This lectin family has a characteristic sequence motif in the CRD and a novel structural fold. The expansion of tandem CRD repeats is also a common observation in this family. In teleost fish, multiple F-lectin isoforms have been shown to be inducible upon inflammatory challenge. Since fucose and their derivatives are also present on the surface of microbial pathogens, it has been suggested that these lectins have a role as recognition factors in innate immune functions (Cammarata et al. (2007)).

Ficolins are humoral proteins in innate immune systems which recognize carbohydrates on pathogens, apoptotic and necrotic cells (Zhang and Ali (2008)). They are structurally similar to mannan-binding lectins (MBLs) due to its collagen-like stalk. Three types of ficolins have been characterized, ficolin-H, -L and -M, which consist of subunits with a short N-terminal domain, a middle collagen-like domain and a C-terminal fibrinogen-like domain. Although these proteins have been found in various species of mammals, they have some differences in pattern of expression, localization, ligand-binding specificity, and complex-formation with MBL-associated serine proteases (MASPs). The ficolins that can form complexes with MASPs and small MBL-associated proteins (sMAPs) activate the complement system through what is called the lectin pathway (Endo et al. (2007)). Although they are claimed to be a subfamily of collectins by some (Runza et al. (2008)) because of their presence on plasma and mucosal surfaces and their similar functions in the immune system, they are actually structurally different: the CRDs of collectins are C-type lectin domains whereas ficolins are fibrinogen-like (Holmskov et al. (2003)). MBLs are also known to form complexes with MASP, with MASP-2 being the main initiator of the lectin complement pathway (Garred (2008)).

Galectins constitute the most widely occurring family of animal lectins, consisting of structurally homologous β-galactoside-binding proteins having a tertiary structure that is similar to both legume lectins and vetebrate pentraxins. There are twelve mammalian galectins, numbered galectin-1, galectin-2, etc. and many more in other species such as birds, lower vertebrates, worms and sponges (Leffler (2001); Rabinovich et al. (2002); Vasta et al. (1997)). They occur in almost all types of cells, within and without, with a tendency to be found more profusely in specific cell types. For example, galectin-4 and -6 are almost exclusively found in epithelial cells, whereas galectin-5 is found in erythrocytes and galectin-7 is exclusive to keratinocytes (Sharon and Lis (2007)). Although galectin-1 and galectin-3 knock-out mice have shown a virtually normal phenotype, there is substantial evidence that galectins have a wide array of functions. For example, it has been shown that they play a role in cell adhesion and immunity regulation (Rabinovich et al. (2007)). It has also been proposed that galectins function as scaffolding proteins on the cell surface to organize cell-surface glycoproteins into functional domains in the plasma membrane (Garner and Baum (2008)). Consequently, it makes sense that galectins may play a role in cancer; it has been shown that galectin-3 promotes the spread of colon and other cancer cells (Gabius (2008)). They have also been shown to have immunoregulatory roles in intestinal inflammatory disorders. Galectin-1 and galectin-2 have shown to contribute to the suppression of intestinal inflammation by inducing apoptosis of activated T cells, whereas galectin-4 was shown to exacerbate this inflammation by stimulating intestinal CD4+ T cells to produce IL-6 (Hokama et al. (2008)). Thus it is suggested that galectins may be a therapeutic target or used as therapeutic agents for inflammatory diseases, cancers, etc. (Yang et al. (2008)).

I-type lectins are lectins having an immunoglobulin (Ig)-like domain, capable of recognizing sialic acids, all kinds of N-glycans, and glycosaminoglycans. I-type lectins other than siglecs include cell adhesion molecule L1, neural cell adhesion molecule (NCAM), myelin protein zero (P_0 or MPZ), and intercellular adhesion molecule-1 (ICAM-1). L1 recognizes α2-3 linked sialic acids on CD24, NCAM recognizes high-mannose N-linked glycans, P_0 recognizes hybrid and complex-type N-glycans, and ICAM-1 recognizes hyaluronan. I-type lectins also include proteins that recognize sulfated GAGs. For example, fibroblast growth factor receptors (FGFRs) and perlecan have several Ig-like domains, and they are both known to interact with heparin/heparan sulfate (Angata and der Linden (2002)). Siglecs (sialic acid-binding Ig-like lectins) constitute the major homologous subfamily of I-type lectins and can be divided into two groups based on evolutionary conservation: Siglecs-1, -2 and -4 form one group, while Siglecs-3 and -5-13 in primates form the second. All siglecs are single type-I integral membrane proteins containing extracellular domains with at least one unique and homologous N-terminal V-set Ig domain, followed by variable numbers of C2-set Ig domains. They contain a conserved Arg residue which is known to form a salt bridge with the carboxylate which is required in most cases for glycan binding. Siglecs are known to be functional in the hematopoietic and immune cell systems in human, having a wide diversity of specificity for different forms of sialylated glycan structures, as shown in Figure 2.12 (Varki and Angata (2006)).

X-lectins were named after the *Xenopus laevis* oocyte cortical granule lectin XL35, based on which many homologous proteins have been found in frog, human, mouse, lamprey, trout, and ascidian worm (Lee et al. (2004)). It was found that the proteins in *Xenopus* exhibited multiple functions as their expression patterns varied during development. The mammalian homologues of X-lectins are called intelectins, which have been found in intestinal tissues, and function in innate immunity. Thus it has been suspected that X-lectins also function as defense proteins (Ishino et al. (2007)).

L-type lectins are distinguished from other lectins primarily based on tertiary structure. In general, the structure contains antiparallel β-sheets connected by short loops and β-bends and no α-helices. The entire structure forms a jelly-roll fold, which may also be called a lectin fold. L-type lectins were first discovered in plants, but their biological role is still unclear. One hypothesis is that they serve as storage proteins to nourish the plant. Another is that they have a role in plant defense, being toxic to some insects, or as pattern-recognition receptors within the plant innate immune system (Varki et al. (2008)). Calnexin and Calreticulin are homologous lectin-like molecular chaperones with L-type lectin domains that interact with newly synthesized glycoproteins in the ER. Both are monomeric, calcium-binding proteins, related to members of the legume lectin family. During N-glycosylated protein biosynthesis, after αglucosidases I and II remove two glucose residues, if the

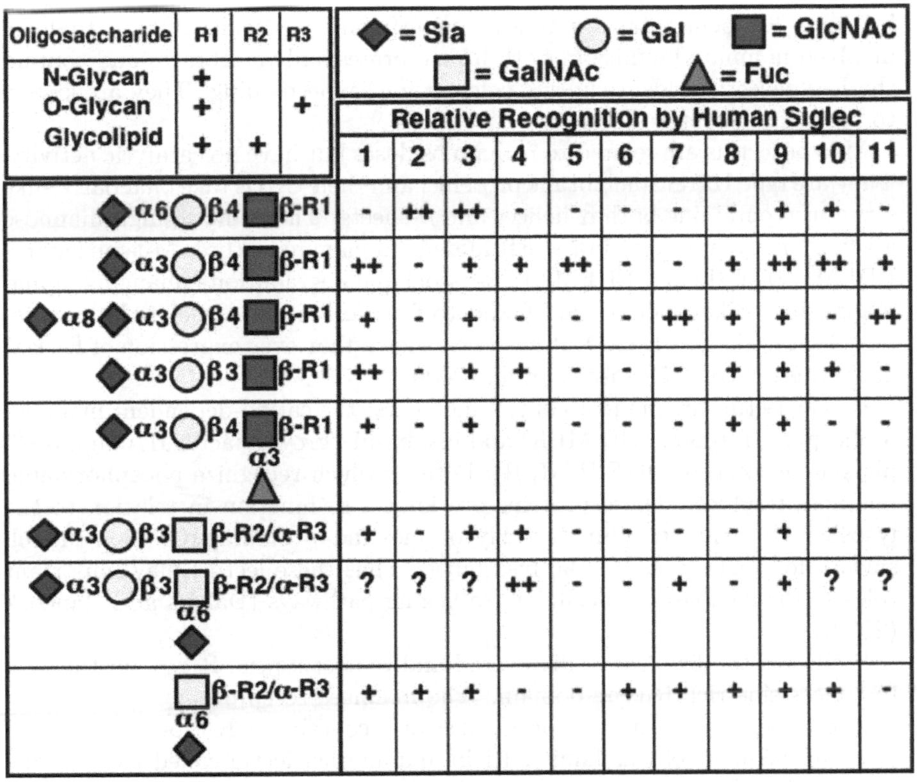

Oligosaccharide	R1	R2	R3	◆ = Sia ○ = Gal ■ = GlcNAc □ = GalNAc ▲ = Fuc										
N-Glycan	+			**Relative Recognition by Human Siglec**										
O-Glycan	+		+	1	2	3	4	5	6	7	8	9	10	11
Glycolipid	+	+												
◆α6○β4■β-R1				-	++	++	-	-	-	-	-	+	+	-
◆α3○β4■β-R1				++	-	+	+	++	-	-	+	++	++	+
◆α8◆α3○β4■β-R1				+	-	+	-	-	-	++	+	+	-	++
◆α3○β3■β-R1				++	-	+	+	-	-	-	+	+	+	-
◆α3○β4■β-R1 (α3 ▲)				+	-	+	-	-	-	-	+	+	-	-
◆α3○β3□β-R2/α-R3				+	-	+	+	-	-	-	-	+	-	-
◆α3○β3□β-R2/α-R3 (α6 ◆)				?	?	?	++	-	-	+	-	+	?	?
□β-R2/α-R3 (α6 ◆)				+	+	+	-	-	+	+	+	+	+	-

FIGURE 2.12: The glycan specificities of human siglecs. *Figure reused by permission of Oxford University Press.*

glycosylated proteins cannot fold properly in time, then calnexin and calreticulin bind to and retain these proteins having Glc1Man9GlcNAc2 structures to facilitate their folding. If proper folding cannot be obtained in time, glucosidase II removes the last glucose from the glycan such that the glycoprotein can be liberated from calnexin/calreticulin (Parodi (2000)). Monoglucosylated glycans are then recreated by the glucosyltransferase such that calnexin/calreticulin can bind to them once again for another chance at proper folding. Alternatively, the misfolded glycoproteins with the Man9GlcNAc2 structure may be acted on by a mannosidase such that it can be retro-translocated out of the ER to the cytosol to be degraded by proteasomes, in what is called ER-associated degradation, or ERAD (Hebert et al. (2005)). ERGIC-53 and VIP36 are type I membrane proteins that have the L-type lectin domain and participate in vesicular protein transport in the secretory system. They both bind to oligomannose-type glycans and require calcium for binding. The pentraxins are another superfamily of plasma proteins that are involved in innate imunity in both invertebrates and vertebrates and contain the L-type lectin folds, requiring calcium for ligand binding. They are known to recognize galactose residues.

M-type lectins are related to α-mannosidases but have no catalytic activity. They are type II transmembrane proteins, and their CRDs take on a barrel-like structure containing both α-helices and β-sheets. They bind to high-mannose glycans on glycoproteins in the ER lumen. In mammals, three M-type lectins, EDEM1, EDEM2 and EDEM3 (ER-associated degradation-enhancing alpha-mannosidase-like proteins) are found to function in ER-associated glycoprotein degradation, or ERAD. It has been shown that over-expression of EDEM accelerates ERAD (Hosokawa et al. (2006)).

P-type lectins are made up of two members: the cation-dependent mannose 6-phosphate receptor (CD-MPR) and insulin-like growth factor II/mannose 6-phosphate receptor (IGF-II/MPR), both of which recognize phosphorylated mannose residues. These proteins are known to function in relation to the lysosomes by generating functional lysosomes and by binding IGF-II at the cell surface for degradation in the lysosomes. Thus these lectins play important roles in the complex intracellular trafficking pathways (Dahms and Hancock (2002)).

Ricin was the first lectin discovered and represents the R-type lectins due to its cysteine-rich R-type domain. The mannose receptor subfamily, which is also a member of the C-type lectins, also contain an R-type domain and are thus members of this family. Ricin and another lectin called RCA-I were first purified from *R. communis* seeds. Ricin is an agglutinin as well as a very potent toxin. It binds to β-linked galactose and GalNAc, whereas RCA-I prefers β-linked galactose only. In general, ricin and RCA-I both preferentially bind to Galβ1-4GlcNAc or GalNAcβ1-4GlcNAc, and they have weak binding to Galβ1-3GlcNAc. However, binding affinities are rather low, despite their high affinities in cell-binding, indicating that these lectins are multivalent (Varki et al. (2008)).

Tachylectins have been identified in the Japanese horseshoe crab, *Tachypleus tridentatus* and consist of five members numbered Tachylectin-1 through Tachylectin-5. Tachylectin-1 binds to LPS as well as polysaccharides such as agarose and dextran. Tachylectin-2 binds D-GlcNAc and D-GalNAc as well as LPS. Tachylectins-3 and -4 specifically bind to S-type LPS from Gram-negative bacteria through a certain sugar moiety on O-antigens. Finally, Tachylectin-5 recognizes acetyl-group-containing substances which may or may not be glycans. Tachylectin-5 contains a short N-terminal cysteine-containing segment and a fibrinogen-like domain in the C-terminal, with more than 50% sequence identity to mammalian ficolins. However, it lacks the collagen domain of ficolins. It is thought that these lectins are used to recognize invading pathogens in the innate immune system of the horseshoe crab (Kawabata and Iwanaga (1999); Gokudan et al. (1999); Beisel et al. (1999)).

Leczymes are lectins that also have enzymatic activities (Nitta (2001)). In fact, ricin of the R-type lectins is a leczyme, having a chain that functions as an RNA N-glycosidase (Varki et al. (2008)). Other leczymes include the sialic acid-binding lectin from *Rana catesbeiana*, which has ribonuclease activity (Iwama et al. (2001)), and the LNP lectin from *Dolichos biflorus* which acts as an apyrase that catalyzes the hydrolysis of phosphoanhydride bonds in nucleosides of di- and tri-phosphates (Roberts et al. (1999)).

2.2.2 Carbohydrate-binding mechanism of lectins

The carbohydrate-binding mechanism of lectins is known to be mediated by the network of hydrogen bonds and hydrophobic interactions that can form between them. In rare cases, electrostatic interactions, or ion pairing, as well as metal ion coordination may also affect binding. Water molecules may also mediate binding, and on the whole, a variety of amino acids may participate. In particular, Asp, Asn, Glu, Gln, and Arg on the side chains of proteins often interact with the hydroxyl groups (OH) of glycans. The numerous hydroxyl groups on glycans facilitates the formation of multiple hydrogen bonds with different amino acids in the binding site of a lectin since these OH groups can both donate and accept hydrogen bonds simultaneously.

It must be noted here that most lectin-monosaccharide interactions are relatively weak, and that strong binding occurs for oligosaccharides of cell surface glycoproteins and glycolipids, suggesting that multiple protein-carbohydrate interactions actually occur simultaneously. There are four possible ways in which such multivalent binding can occur: (a) ligand multivalency, (b) an extended binding region capable of interacting with more than one monosaccharide on a glycan, (c) clustering of several identical binding sites by forming protein oligomers that can simultaneously bind different glycans spaced appropriately, and (d) a combination of (a) and (c). For example, concanavalin A has shown increased affinity for a synthetic polymer of multiple mannose residues compared to methyl α-mannoside (Mortell et al. (1996)). For example, it has been shown that the affinity of human mannose receptor increased with a series of lysine-based cluster mannosides when the number of mannose residues per molecule were increased from two to six (Biessen et al. (1996)). As another example, the starburst glycodendrimers are tree-shaped molecules with carbohydrates at the outer periphery (Roy (1996)). These highly dense balls of carbohydrates enhance affinities to lectins by several orders of magnitude (Sharon and Lis (2007)).

Multivalency has shown to affect the specificity of lectins as well. For example, whereas concanavalin A binds MeαMan with a four-fold higher affinity than MeαGlc, the polyvalent derivatives of these monosaccharides resulted in an up to 160 fold difference in affinity (Mortell et al. (1996)). Similarly, a *Bauhinia purpurea* lectin was shown to prefer Galβ1-3GalNAcβ over Galα1-3GlcNAcα in solution, but then switched its preference when they were immobilized (Horan et al. (1999)). Multivalency also enables the formation of diverse arrays of oligosaccharides and lectins (Brewer (1996)). For example, linear arrays can be obtained from divalent oligosaccharides bound to dimeric lectins, end-to-end. Three-dimensional lattices may also be formed when tetrameric lectins are bound to divalent oligosaccharides (Sacchettini et al. (2001)).

2.3   Carbohydrate-carbohydrate interactions

Up to this point, this section has introduced the wide variety of carbohydrate-protein interactions that may take place in biological systems. However, the interactions between carbohydrates themselves are also quite important in cell biology in that they offer a rich supply of potential low-affinity binding sites on cell surfaces, which may be arranged in a polyvalent array that creates a flexible and versatile carbohydrate-carbohydrate recognition system. This carbohydrate recognition process may be based on (1) self-recognition of carbohydrates on interfacing cell surfaces or (2) recognition of different carbohydrates from oneself on different types of cells (Figure 2.13).

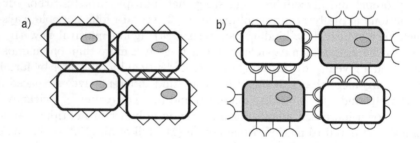

FIGURE 2.13:   Examples of carbohydrate-carbohydrate interactions. a) Self-recognizing carbohydrates on homogeneous cells. b) Recognition of different carbohydrates on different cells.

One of the first examples of carbohydrate-carbohydrate interactions were identified in sponges, which contained membrane adhesion molecules responsible for calcium-dependent cell-cell recognition. These were large proteoglycans whose structures were 30-60% carbohydrates. Atomic force microscopy visualization of these proteoglycans revealed a linear structure and a sunburst-like core structure with 20-25 radiating arms. In the sunburst-like structure, two types of *N*-linked glycans were found: the larger glycan was found in the core and the smaller were found in the arms. It is surmised that the arms bind to the cell surface through a calcium-dependent protein-carbohydrate interaction on the cell surface receptor and that the core structure interacts with other core structures on other cells in order for the cell-cell interaction to take place.

In another example, mouse lymphoma cells expressing gangliotriaosylceramide (Gg3) and melanoma cells expressing sialosyllactosylceramide (GM3) were found to aggregate based on the interaction of these expressed gangliosides. Gg3 corresponds to the structure NeuAcα2-3Galβ1-4Glcβ1-Cer, and GM3 is GalNAcβ1-4Galβ1-4Glcβ1-Cer. The GM3-dependent adhesion of melanoma cells to endothelial cells increases the motility of the melanoma cells, thus promoting the advancement of melanoma cell metastasis (Bucior and Burger (2004)).

Chapter 3

Databases

The impetus behind the development of glycome informatics was the construction of large-scale databases for storing a comprehensive dataset of glycan structure data. Originally, the Complex Carbohydrate Structure Database (CCSD) (Doubet et al. (1989)) was developed at the Complex Carbohydrate Research Center at the University of Georgia in the 1990's. This database became better known as CarbBank, which was the name of the tool to perform queries on CCSD. The database was discontinued in the mid-90's, but the data was still made available to the public and thus became the foundation for other glycan structure databases that followed. In terms of glycome informatics, the first group to develop computer-theory-based algorithms for carbohydrate structures was KEGG. Before then, GLYCOSCIENCES.de had accumulated carbohydrate data from both CarbBank and PDB, along with experimental information such as mass spectra. Furthermore, the Consortium for Functional Glycomics (CFG) started their own database of glycan structures to be associated with their experimental data as well.

In addition to glycan structures, the analysis of glycans often involves other types of molecules such as lipids and glycan-binding proteins. Therefore, this chapter will introduce not only the major glycan structure databases publicly available, but also some glycan-related databases that may be useful for bioinformatics research.

3.1   Glycan structure databases

As mentioned above, the major glycan structure databases that are publicly available are KEGG GLYCAN, GLYCOSCIENCES.de, and that developed by the Consortium for Functional Glycomics (CFG). Other databases include the Bacterial Carbohydrate Structure DataBase (BCSDB), which is a comprehensive database of carbohydrate structures found in bacteria, and GLYCO3D, which is a database of three-dimensional structures of glycans and related proteins, mainly extracted from PDB. Additionally, GlycomeDB is the newest addition to the glycan structure databases, containing an integrated database of all glycan structures from KEGG GLYCAN, GLYCO-SCIENCES.de, CFG, BCSDB, and others. GlycomeDB was developed out of the EuroCarbDB project, which was a project aimed towards creating a foundation for databases and bioinformatics tools in glycobiology and glycomics. This project also produced a database called MonosaccharideDB, which is a database of monosaccharides, constructed for the purpose of standardizing the nomenclature of monosaccharides used by glycan structure databases. Each of these databases will be described in this section.

3.1.1   KEGG GLYCAN

URL http://www.genome.jp/kegg/glycan/

FTP ftp://ftp.genome.jp/pub/kegg/ligand/glycan

The KEGG GLYCAN database (Hashimoto et al. (2006)) was developed as a new database under the KEGG LIGAND database of chemical compound structures in the Kyoto Encyclopedia of Genes and Genomes (KEGG) resource (Kanehisa et al. (2008)). The structures were originally derived from the CarbBank database. The KEGG group took the structures in Carb-Bank and found that there were many duplicated structures. Thus, they used a tree-structure alignment algorithm (KCaM, described in Section 4.2.1) to find exact-matching structures to combine them into single glycan structure entries. Each unique entry was then given an ID beginning with the letter "G" followed by five digits, in line with the ID system for all databases in KEGG. As a result, over 40,000 entries in CarbBank were consolidated into almost 11,000 unique glycan structures.

3.1.1.1   Database content

The main entry point for the KEGG GLYCAN database contains a number of links to tools and information regarding the glycan structures available, described in Table 3.1. CSM stands for Composite Structure Map, which is a tool to display related glycan structures from a global perspective. Users have the option to select a common core structure for the glycans to view. Then those glycans that differ by a single glycosidic linkage are connected to one another in a global map. Thus nodes represented by monosaccharides actually represent a whole glycan structure and edges represent enzymes that synthesize or remove the linkage. An example is given in Figure 3.1, where monosaccharides represent the glycan structure containing the linear substructure along the path from the selected monosaccharide up to the root node. Clicking on a monosaccharide symbol will display all the glycan structures containing the chain of sugars from the root on the right up to the selected node. Moreover, clicking on a red link will display the enzyme (either species-specific or the ortholog group, depending on the selected view from the pull-down menu at the top) that synthesizes and/or removes the selected linkage.

The KegDraw and KCaM tools are described in Section 3.1.1.2. The GECS (Gene Expression to Chemical Structure) tool is also described in detail in

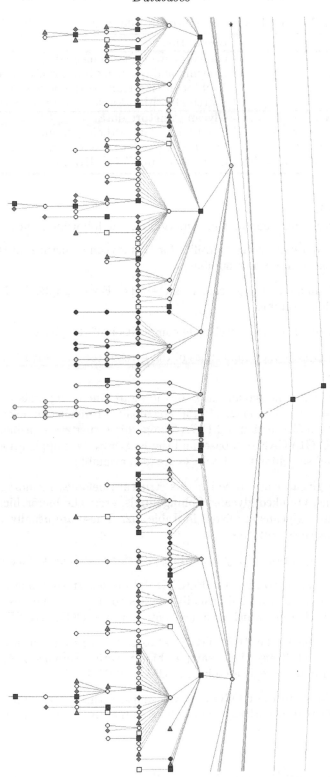

FIGURE 3.1: Snapshot of the CSM for the *N*-glycan core up to the first mannose.

TABLE 3.1:   Tools and information available in KEGG GLYCAN

Name	Description
CSM	Global view of all "connected" glycan structures
KegDraw	Drawing and querying tool for glycan structures
GECS	Tool for linking transcriptomic data to glycan structure
KCaM	Search algorithm for similar glycan structures
GLYCAN	Glycan structure data
Glycosyltransferases	Categorized into ortholog groups
Pathway maps	Metabolic, regulatory and structure maps
Glycan binding proteins	Categorized in KEGG BRITE

Section 4.3.1.

Each glycan entry contains information for the following items:

Entry   The database entry identifier for this glycan structure (in the format *Gnnnnn* where n is a number).

Name   The name of the glycan, if one exists. For example, LacNac for N-acetyllactosamine.

Composition   The monosaccharide composition of the glycan.

Mass   The average molecular mass of the glycan, computed from the composition.

Structure   A 2-dimensional image of the figure, including links to the KCF format (see Section 2.1.4) for download. Buttons for executing the KCaM tool (Section 4.2.1) for using this structure as a query to the KEGG GLYCAN database, and the KegDraw Java application for editing and searching the database are also available.

Class   The glycan class to which this structure belongs, including N-linked glycans, O-linked glycans, Sphingolipids, etc. The hierarchical classification of glycans are listed in Table 3.2. These are usually written as *Glycan class; Subclass.*

Binding   Information, if any, regarding proteins that bind to this glycan.

Compound   Links to the COMPOUND database corresponding to this glycan entry, if any. This usually applies to monosaccharides and small oligosaccharides which also have entries in the COMPOUND database.

Reaction   Links to the REACTION database to entries in which this glycan is involved, if any. For example, biosynthetic reactions producing this glycan may be registered in the REACTION database.

Pathway   Links to the KEGG pathway maps, if this glycan appears in any. The displayed pathway map will have the glycan entry highlighted in red.

TABLE 3.2: Hierarchical classification
of glycans in KEGG GLYCAN

Class	Subclass
Glycoprotein	N-Glycan
	O-Glycan
	Glycosaminoglycan
	GPI Anchor
	Others
Glycolipid	Sphingolipid
	Glycerolipid
	LPS
	Others
Polysaccharide	N/A
Oligosaccharide	N/A
Glycoside	N/A
Neoglycoconjugate	N/A
Others	N/A

Enzyme Similar to the Reaction field, where links to the ENZYME database
are made available if this glycan is involved in any enzymatic reaction
entries in the ENZYME database.

Ortholog Links to the KEGG Orthology database entries corresponding to
the enzymes listed in the Enzyme field.

Reference Literary references citing this glycan structure.

Other DBs Links to other databases outside of KEGG which provide the
same glycan structure. This field usually contains links to the CarbBank
(CCSD) database.

LinkDB Links to the LinkDB system, which automatically generates links
to other database entries in DBGET, including such databases as Swiss-
Prot and RefSeq.

KCF The KCF representation of the glycan structure. This can be used
for drawing and editing glycan structures in the KegDraw tool and for
querying the database, as described in the next section. Refer to Section 2.1.4 for details on the KCF format.

In addition to the glycan structure data in KEGG GLYCAN, KEGG also
includes a number of glycan-related data, including hierarchical classifications of glycan binding proteins and glycosyltransferases, organized in KEGG
BRITE. The KEGG Pathways also include metabolic and regulatory pathways as well as Glycan Structure Maps, which display glycosyltransferases on
top of glycan structures, as opposed to wire-diagrams of enzymes. The KEGG
Orthology, or KO, system of orthologous genes is also included in BRITE. All

this information is available from the KEGG GLYCAN entry point at the URL above.

3.1.1.2   Database queries

Queries of the KEGG GLYCAN database can be made in both textual form for searching keywords and structural form for searching for similar glycan structures. For queries by keyword search, users may use the DBGET search engine in KEGG. Queries by keywords such as KEGG GLYCAN identifier, name, or class may be performed. As an example, from the DBGET search engine for KEGG GLYCAN at `http://www.genome.jp/dbget-bin/www_bfind?glycan`, entering "LacNAc" for N-acetyllactosamine will retrieve four glycan structures containing LacNAc.

For structural queries, either KegDraw or the KCaM Search Tool can be used. KegDraw is a downloadable Java application which can be used to make queries across the Internet. The ChemDraw-like interface makes drawing glycan structures straightforward with the mouse. Details on its usage are described in Section 4.5.1.2

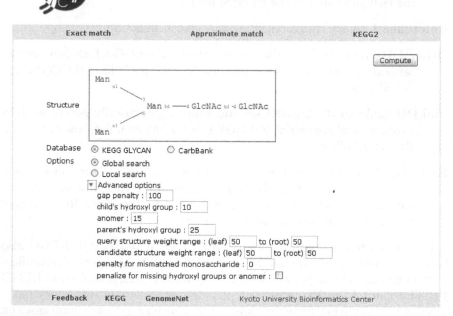

FIGURE 3.2: KEGG GLYCAN search tool initialized with the query structure as inputted in KegDraw and with advanced options shown.

Glycan structures can be saved in KCF format, and they can also be exported in LINUCS format (see Section 2.1.5) or as an image file in PNG format. To use the drawn structure as a query to the KEGG GLYCAN database, the "Search Similar Structures" menu option under the "Tools" menu can be used. When a search is initiated, a web browser window will be displayed with the query structure inputted, ready to be executed from the KEGG Glycan Search Tool.

Advanced options can be selected by clicking on the "Show advanced options" link, as in Figure 3.2. From this screen, a variety of options can be specified, as described below.

1. **Database:** Select the database against which to perform the query.

   (a) **KEGG GLYCAN:** Search the KEGG GLYCAN database.

   (b) **CarbBank:** Search the original CarbBank database.

2. **Program:** Select the type of search to perform.

   (a) **Gapped (Approximate match):** Search for similar structures, specifying a gap penalty value.

   (b) **Ungapped (Exact match):** Search for structures exactly like the query.

3. **Option:** Specify the options for the search program.

   (a) **Global search:** When gapped search is selected, this option will try to make as many matches as possible irregardless of the number of gaps. When ungapped match is selected, as many exact matches as possible will be found.

   (b) **Local search:** When gapped match is selected, this option will try to minimize the number of gaps whereas when ungapped match is selected, the maximal exact match will be returned.

4. **Advanced options:** For advanced users, the default scoring parameters used for the search programs can be adjusted.

   (a) **child's hydroxyl group:** A score can be specified for matching hydroxyl groups of children (closer to the non-reducing end).

   (b) **anomer:** A score can be specified for matching anomeric information.

   (c) **parent's hydroxyl group:** A score can be specified for matching the hydroxyl group of the parent (closer to the reducing end).

   (d) **query structure weight range:** A gradient for scoring components based on their position in the query structure can be specified by entering values for the extremes at the leaf (non-reducing end) and the root (reducing end).

(e) **candidate structure weight range:** A gradient for scoring components based on their position in the database structure can be specified by entering values for the extremes at the leaf (non-reducing end) and the root (reducing end).

(f) **penalty for mismatched monosaccharide:** a penalty value for mismatched monosaccharides can be specified such as to modulate the acceptance of gaps.

(g) **penalize for missing hydroxyl groups or anomer:** by checking this box, missing glycosidic bond information will either be penalized or ignored.

The results of the query in Figure 3.2 are displayed in Figure 3.3. An image of each structure is given, along with the corresponding similarity score, composition, and class information. For the default parameters, the similarity score is given as the number of matched components times 100. That is, for the approximate match, if five monosaccharides exactly match the query, then the maximum score possible is 500. Furthermore, clicking on the score will display a window illustrating the alignment of the query with the selected structure.

3.1.2 GLYCOSCIENCES.de

URL http://www.glycosciences.de/

GLYCOSCIENCES.de is a portal for glycomics research, containing not only carbohydrate structure data, but also tools for glycomics analysis (Lutteke et al. (2005)). In particular, special emphasis has been placed on the availability of experimentally determined structures in 3D space and their interactions with proteins. The tools provided for glycomics analysis are listed in Table 3.3.

3.1.2.1   Database content

The GLYCOSCIENCES.de database provides a wide variety of information for carbohydrates, listed below, which can each be searched through their individual interfaces. The types of information available in GLYCOSCIENCES.de includes:

2D structure The structure of the glycan in CarbBank format (See Section 2.1.3).

Structure motifs Information regarding known glycan motifs found in the given entry.

Chemical information The chemical formula, molecular weight, number of atoms, residues, etc.

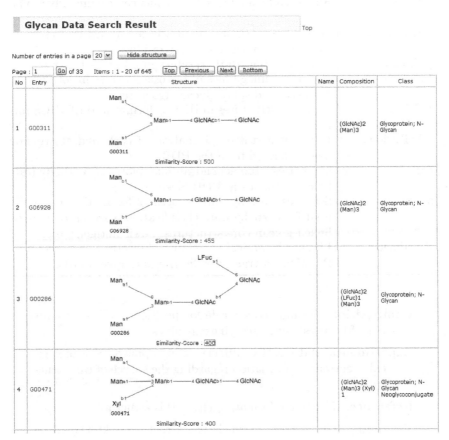

FIGURE 3.3: KEGG GLYCAN search results from an approximate search of the *N*-linked glycan core structure.

TABLE 3.3:   Tools and information available in GLYCOSCIENCES.de

Name	Description
GlycoFragment	Calculates and displays the main fragments (B- and C-, Z- and Y-, A- and X-ions) of oligosaccharides that should occur in MS spectra.
pdb2linucs	Automatically extracts carbohydrate information from pdb-files and displays it in LINUCS format.
GlySeq	Statistically analyzes the sequences around glycosylation sites.
GlyVicinity	Generates statistics about the amino acids present in the vicinity of carbohydrate residues.
pdb-care	Checks carbohydrate residues in pdb-files for errors.
LiGraph	Generates image files of oligosaccharides in formats often used to display glycan structures.
LINUCS	Converts structures in IUPAC format into LINUCS format.
GlyTorsion	Performs a statistical analysis of carbohydrate torsion angles derived from the PDB.
carp	Generates Ramachandran-like plots of carbohydrate linkage torsions in PDB files.
sumo	Searches carbohydrate structures for motifs commonly used for carbohydrate classification, such as *N-* and *O-*linked glycan core structures, Lewis antigens, etc.
PubFinder	Finds thematically related literature by scanning PubMed abstracts for discriminating keywords.

Composition The monosaccharide composition, such as the number of hexoses and N-acetylhexosamines.

Experimental data NMR, MS, crystallographic data or biological occurrence information regarding the source of this structure.

References Literary information citing this structure.

Taxonomy The species in which this structure is known to be found.

3.1.2.2   Database queries

In addition to structure search, queries for and using other types of data are also available. Literary references can be searched by author or title, and nuclear magnetic resonance (NMR) and mass spectrometry (MS) can be searched using peak lists, for example. Furthermore, PDB data can be queried by a variety of options, and searches directly by LINUCS ID are also available.

There are several options to retrieve glycan structure data in GLYCO-SCIENCES.de. The exact match search can be accomplished using any of several formats, including LINUCS (see Section 2.1.5), IUPAC (see Section 2.1.2), or CarbBank (see Section 2.1.3) formats. There is also an option to retrieve glycans based on the composition of the residues. This is especially useful for searching for structures whose specific residues of the same mass cannot be distinguished. A range for the number of residues, such as hexoses, can thus be queried with this option. An example is given in Figure 3.4, where a query for structures containing two sialic acid residues, at least four hexose residues, and at least two HexNAc residues have been specified.

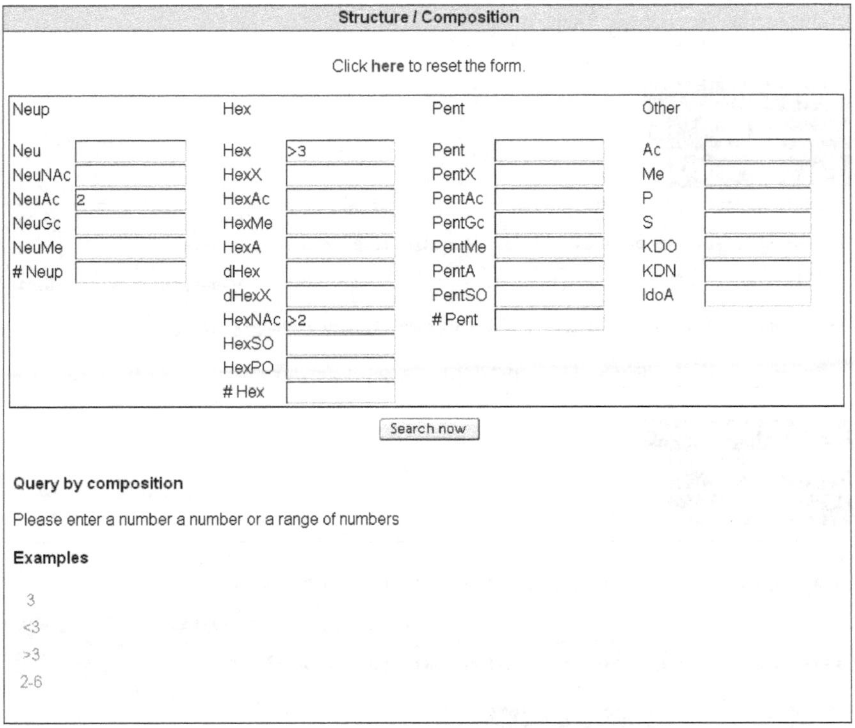

FIGURE 3.4: GLYCOSCIENCES.de composition search tool, where a form for specifying the number of different types of residues can be entered, useful for when the specific monosaccharides are unknown.

The results of this query are displayed in Figure 3.5. Each matching entry is given in CarbBank format along with its composition. There are also buttons for further analysis, including the *Explore* and *LiGraph* buttons. The *Explore* button will display further detailed information regarding the selected structure, as described above. The *LiGraph* button will display the LiGraph tool (introduced in Section 4.5.1.1), which can convert the selected structure to other formats, including IUPAC or an image file using notations such as the CFG representation (see Section 1.2).

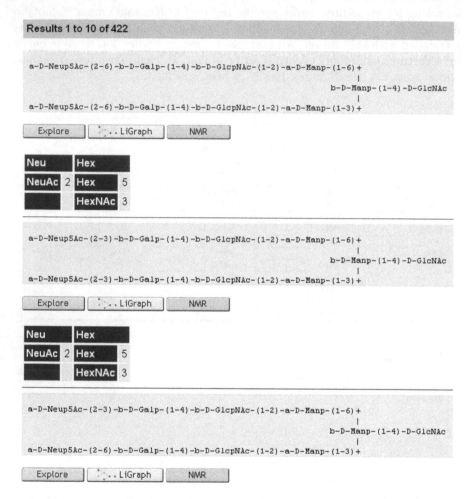

FIGURE 3.5: GLYCOSCIENCES.de composition search results of the query in Figure 3.4.

For substructure search in "beginner" mode, users are provided with a form where entries corresponding to glycosidic components can be entered, as in Figure 3.6. The "advanced" mode allows users to specify two "beginner" mode structures that can be logically combined using AND or OR. Since most entries contain on average five to ten residues, these modes are considered to be sufficient for almost all queries.

Another useful tool is the Motif Search, which includes well-known glycan

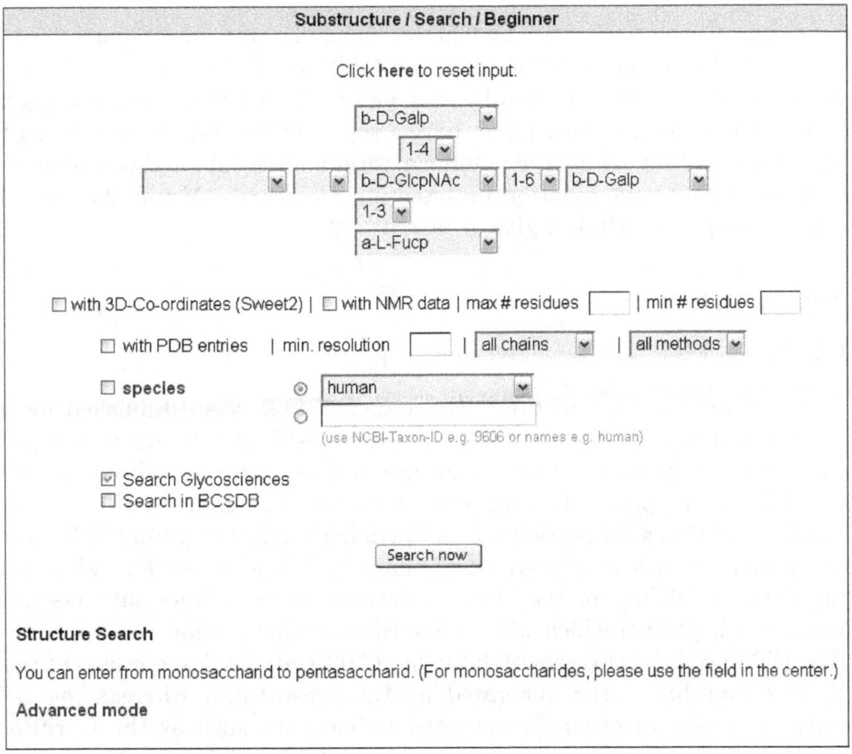

FIGURE 3.6: GLYCOSCIENCES.de beginner search tool, where a form for specifying residues and glycosidic linkages to be searched can be entered.

TABLE 3.4:   CFG Data, provided by the analytical services of the CFG.

Data	Description
Glycan Profiling	MS data of *N*- and *O*-linked glycans in human and mouse tissues as well as various cell lines
Gene Microarray	Screening of RNA samples provided by investigators against a microarray chip containing the latest glycogenes
Glycan Array	Lectin-ligand interaction via glycan arrays

substructures such as Lewis^x and blood group H antigen. Figure 3.7 is a snapshot of the structures containing the GM2 motif, for example. *O*-linked glycan core structures may also be searched with this tool. *N*-linked glycan core structures, on the other hand, have a separate interface for searching for their three subtypes, number of antennae, terminal residues and core structure properties such as fucosylation or bisecting GlcNAcs. Motifs can also be specified using this *N*-linked glycan search tool.

3.1.3   CFG

URL http://www.functionalglycomics.org/

The Consortium for Functional Glycomics (CFG) was established by the National Institute of General Medical Sciences (NIGMS) "to define the paradigms by which protein-carbohydrate interactions mediate cell communication" (CFG home page). During their grant period, hundreds of researchers around the world could participate as Participating Investigators (PIs) to request resources such as glycan array analysis, MS analyses and glyco-gene array data. In doing so, they have generated many unique data resources (Raman et al. (2006)) which will be described in this section.

The CFG provides two sets of data: *the CFG Data* section consists of those data sets that have been generated by the consortium, whereas *the CFG Databases* consist of generally accepted public data such as the CarbBank structures. Since both types of data sets are useful for bioinformatics analysis, they will both be described here.

3.1.3.1   CFG Data

The CFG provides analytical services as listed in Table 3.4. The data generated by these services are made available in this *CFG Data* section. The glycan profiling area contains *N*- and *O*-glycan data from matrix-assisted laser desorption ionization (MALDI) MS experiments performed on mouse and human tissue samples. For mouse, the glycan profiles of both wild-type and knock-out genes are provided for such organs as the brain, heart, ovaries and lymph nodes. Similarly, for human, glycan profiles from selected tissues of a number of patients are available, such as the skin, pancreas, liver and

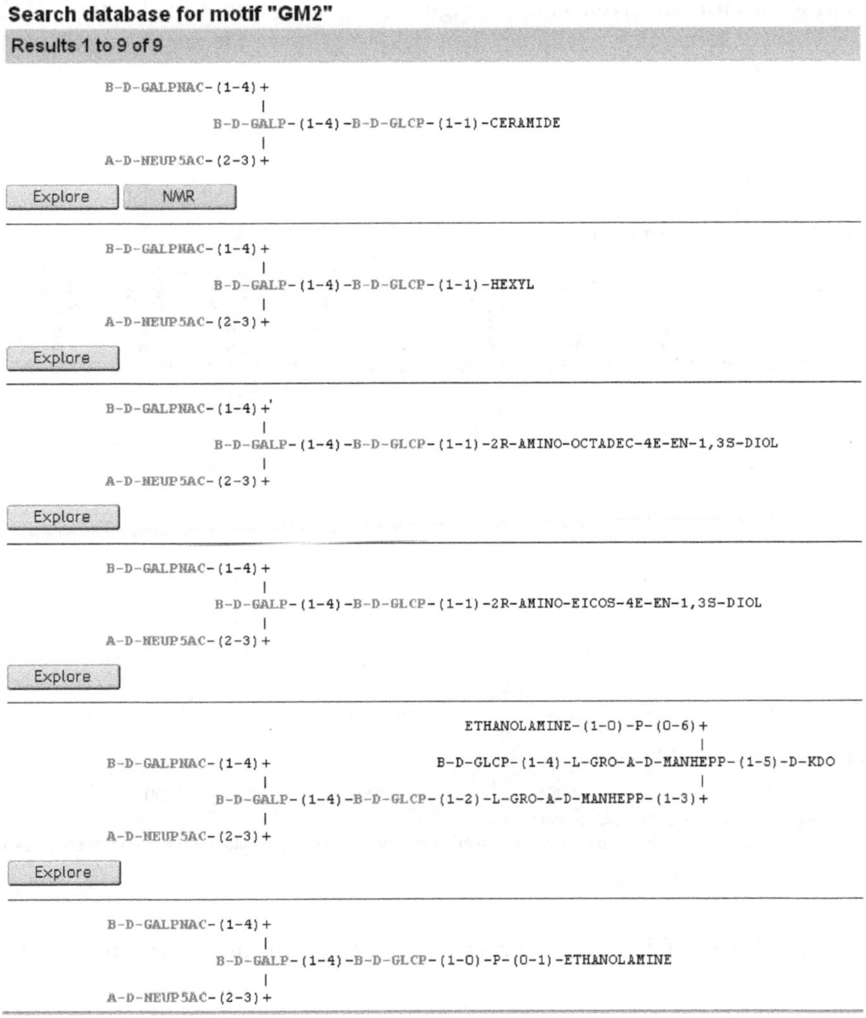

Search database for motif "GM2"

Results 1 to 9 of 9

```
B-D-GALPNAC-(1-4)+
                  |
                  B-D-GALP-(1-4)-B-D-GLCP-(1-1)-CERAMIDE
                  |
A-D-NEUP5AC-(2-3)+
```

[Explore]   [NMR]

```
B-D-GALPNAC-(1-4)+
                  |
                  B-D-GALP-(1-4)-B-D-GLCP-(1-1)-HEXYL
                  |
A-D-NEUP5AC-(2-3)+
```

[Explore]

```
B-D-GALPNAC-(1-4)+
                  |
                  B-D-GALP-(1-4)-B-D-GLCP-(1-1)-2R-AMINO-OCTADEC-4E-EN-1,3S-DIOL
                  |
A-D-NEUP5AC-(2-3)+
```

[Explore]

```
B-D-GALPNAC-(1-4)+
                  |
                  B-D-GALP-(1-4)-B-D-GLCP-(1-1)-2R-AMINO-EICOS-4E-EN-1,3S-DIOL
                  |
A-D-NEUP5AC-(2-3)+
```

[Explore]

```
                                      ETHANOLAMINE-(1-O)-P-(O-6)+
                                                               |
B-D-GALPNAC-(1-4)+                    B-D-GLCP-(1-4)-L-GRO-A-D-MANHEPP-(1-5)-D-KDO
                  |                                             |
                  B-D-GALP-(1-4)-B-D-GLCP-(1-2)-L-GRO-A-D-MANHEPP-(1-3)+
                  |
A-D-NEUP5AC-(2-3)+
```

[Explore]

```
B-D-GALPNAC-(1-4)+
                  |
                  B-D-GALP-(1-4)-B-D-GLCP-(1-O)-P-(O-1)-ETHANOLAMINE
                  |
A-D-NEUP5AC-(2-3)+
```

FIGURE 3.7:   GLYCOSCIENCES.de motif search results of all entries containing the GM2 motif.

spleen. This area also includes glycan profiles of selected cell lines and cell populations such as B-cells, T-cells, NK-cells, neutrophils and leukemic cells. For each sample, glycan profile data can be viewed in the form of one or more images (IMG), in PDF format, or in an applet (APP) developed by PARC (Palo Alto Research Center). This applet allows one to analyze the peaks and annotations for the given sample. Figure 3.8 is a snapshot of this tool. If the MS Voyager software is available, then the DATA data file may be downloaded and analyzed as well.

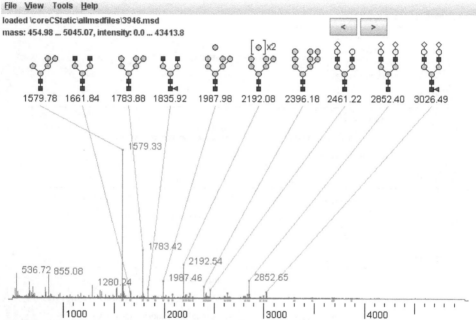

FIGURE 3.8:   CFG applet for analyzing glycan profiling data, provided by PARC.

The gene microarray area contains glyco-gene expression data for RNA samples submitted by participating investigators. The gene chip itself has been continuously updated with the latest glyco-genes. For each experiment, a number of data files, both raw data and processed, are provided. Since each experiment is requested by a participating investigator, after the first period during which the investigator owns the data, it becomes publicly available and listed in the *CFG Data* area. Here, the investigator's name and contact information, the purpose of the microarray experiment, and the samples analyzed are listed. In addition, the gene chip information, and raw and processed data files are also made available. Raw data files can be downloaded in Excel format by clicking the Excel icon, or as raw Affymetrix files by clicking the DAT icon. The data can also be searched by Probe ID or expression value by clicking the spyglass icon. If available, data analysis results are also listed. Under "Low Level" analysis, the CDF file and a compressed file of the Affymetrix analysis can be downloaded. In contrast, under "High Level" analysis, files in either Excel or tab-delimited format can be obtained.

The glycan array area contains glycan-binding affinity data for samples submitted by investigators, which include a wide variety of affinity data such as with lectins and pathogens. A predefined set of glycan structures are spotted on these arrays such that their binding affinities with a variety of other molecules can be measured. Each data sample can be viewed using a graphical interface as displayed in Figure 3.9. By moving the mouse over a particular bar, the glycan structure corresponding to it will be displayed below. Each experiment can also be downloaded as Excel files.

3.1.3.2 CFG Database

The CFG Database section contains information on glycan structures, glycan-binding protein (GBP) molecules and glycosyltransferases. The glycan structure database contains all *N*- and *O*-linked glycan structures from Carb-Bank, data from Glycominds, Ltd., those glycans from the Glycan Profiling area that have been verified, glycan structures spotted on the glycan array and glycans synthesized by the consortium. For a particular carbohydrate entry, as displayed in Figure 3.10, the following information is listed if available:

Glycan Identifier The database entry identifier for this glycan structure.

Cartoon Representation The image of the structure using CFG symbols (see Section 1.2).

IUPAC 2D Representation The image of the structure using IUPAC representation.

IUPAC Code The string representation of the glycan structure in IUPAC format (see Section 2.1.2).

Linear Code The string representation of the glycan structure in Linear Code® format (see Section 2.1.7).

Sub Structure Search Interface A link to perform a substructure search using the glycan.

General Information General information regarding this glycan, such as its class, molecular weight, and composition.

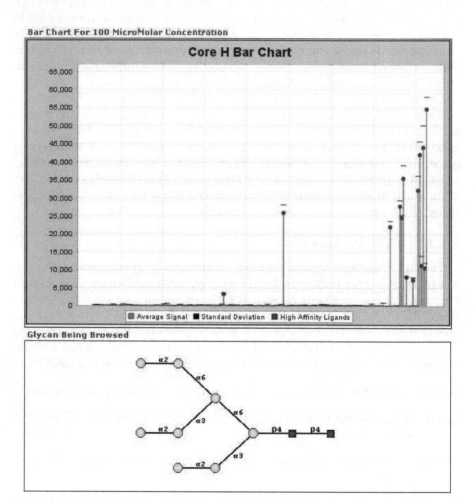

FIGURE 3.9:   The graphical interface for the CFG glycan array data. Moving the mouse over the red bars in the plot will change the display of the selected glycan below.

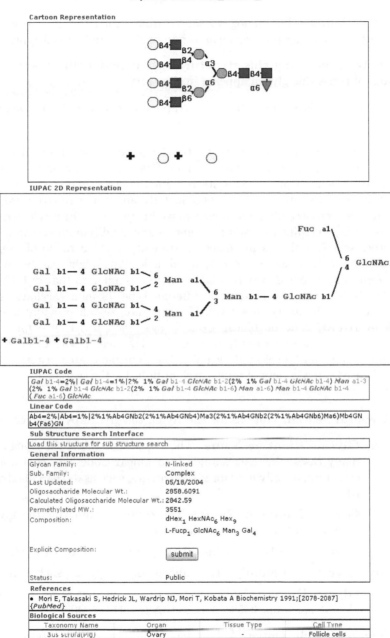

FIGURE 3.10: An example of CFG glycan structure entry ID carbN-link_221394_P.

References Literary references citing this glycan structure.

Biological Sources The biological source from which this glycan was extracted, including the organism, organ or tissue and/or cell type.

PDB entries featuring this structure as a ligand PDB IDs and the number of times this glycan appears in the entry.

Glycosciences.DB Links Links to corresponding entries in GLYCOSCIEN-CES.de.

There are several query options to the CFG glycan structure database. Searches can be made by substructure, molecular weight, composition, linear nomenclature or multiple search criteria. For the substructure search, since the database only contains *N*- and *O*-linked glycans, the core structure with which to start constructing the query must be chosen. Once selected, the structure can be modified by adding, removing or modifying its components.

Another search option is by molecular weight, where a range of weights, either non-permethylated or permethylated molecular weight, can be entered into a form and searched. Composition search is similar to that of GLYCO-SCIENCES.de, where the number of different types of monosaccharides can be specified for the query in a form. Linear Code search provides a string search by IUPAC code or Linear Code® (see Section 2.1.7). Options to search for structures that exactly match, contains, begins with or ends with the input string can be selected. Finally, the multiple search criteria allows one to choose from a variety of text search options for the search, as described below:

1. **General Search:** Options for specifying the glycan by name or class.

    (a) **Glycan Scientific Name:** The scientific name of the glycan. For many cases, this corresponds to a Linear-Code-like name for the structure, usually used for glycan arrays, such as Galb1-3Galb#Sp2. BT.

    (b) **Glycan Common Name:** The common name for the glycan, such as LacNAc for N-acetyllactosamine.

    (c) **Glycan Family:** The class of the glycan, such as N-linked.

    (d) **Glycan Sub Family:** The subclass of the glycan, such as hybrid.

    (e) **Glycan ID:** The CFG identifier for glycan structures. For example, GlycanNlink_35598_A or carbNlink_41417_D000.

2. **Source Search:** Options for the biological source of the glycan.

    (a) **Species:** The species of the glycan in scientific terms (e.g., *Mus* species).

    (b) **Organ:** The organ of the species where the glycan may be found.

(c) **Tissue:** The tissue of the species where the glycan may be found.

(d) **Cell Type:** The cell type from which the glycan was extracted.

(e) **Disease:** Specific disease-related glycans, such as cancer.

3. **Nomenclature Search:** Search by textual representations of glycans using "%" as the wildcard character can be performed.

   (a) **IUPAC Code:** E.g., %Gal%Fuc% can be used to search for structures with Galactose and Fucose residues, with the former closer to the non-reducing end.

   (b) **Linear Code™:** E.g., %A%X% can be used to search for structures with Galactose and Xylose residues, with the former being closer to the non-reducing end.

4. **Oligosaccharide Molecular Weight Search:** A range for the molecular weight (including permethylated) can be specified to limit the size of the results.

   (a) **Oligosaccharide Mol Wt.:** The non-permethylated molecular weight can be specified specifically using the equal ($=$) sign, or a range can be specified by adding a value after the "$+/-$" or by using the less than ($<$) or greater than ($>$) selections.

   (b) **Oligosaccharide Mol Wt. (Permethylated):** Similar to the above, except searches will be performed for permethylated molecular weights.

5. **Composition Search:** Search by composition when the exact topology of the structure is unknown.

   (a) **Monosaccharide types:** A number can be specified for each monosaccharide type.

   (b) **Exact:** If this option is checked, only those structures that exactly match the specified compositions will be returned. If not, then those structure containing at least those specified will be returned.

In all cases, search results include the molecular weight, IUPAC code, composition, family information and the biological source of the matching entries. Clicking on the IUPAC code will display the detailed information regarding the entry, as described earlier.

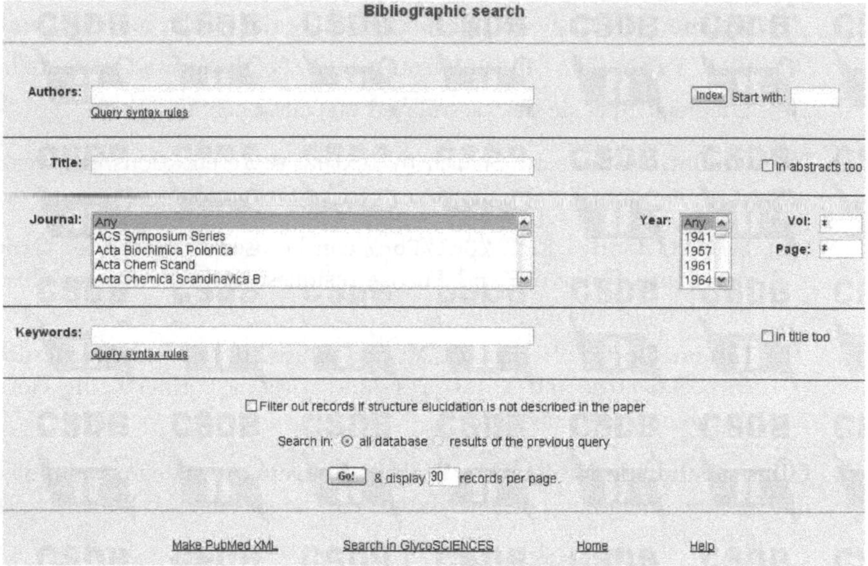

FIGURE 3.11:   The bibliographic search interface of BCSDB. See text for details.

3.1.4   BCSDB

URL http://www.glyco.ac.ru/bcsdb/start.shtml

The Bacterial Carbohydrate Structure DataBase (BCSDB) contains all known carbohydrate structures found in bacteria (Toukach et al. (2005)). There are almost 9000 records of bacterial carbohydrates, which includes approximately 3500 records from CarbBank. The rest of the data has been updated manually from the literature.

Each record in the BCSDB contains the structure, references, abstract of the publications, data on the biological source, methods of structure elucidation, information on the spectral data and assignment of NMR spectra, data on conformation, biological activity, chemical and enzymatic synthesis, biosynthesis, and genetics, among others. Each entry is also linked with the GLYCOSCIENCES.de data, so related information can be retrieved from the more generalized database.

The database can be searched by BCSDB identifier, substructure, NMR spectrum and textual search including microorganism, bibliography and other keywords. For ID search, one or more IDs may be entered in a single text field. Separate IDs can be entered deliminated by commas (,) or ranges of IDs can be specified with a hyphen (-). For example, "3,5,10-12" can be used to search for IDs 3, 5, 10, 11, and 12.

For a bibliographic search, which is illustrated in Figure 3.11, several options are made available. Authors can be searched for by entering a name in the text

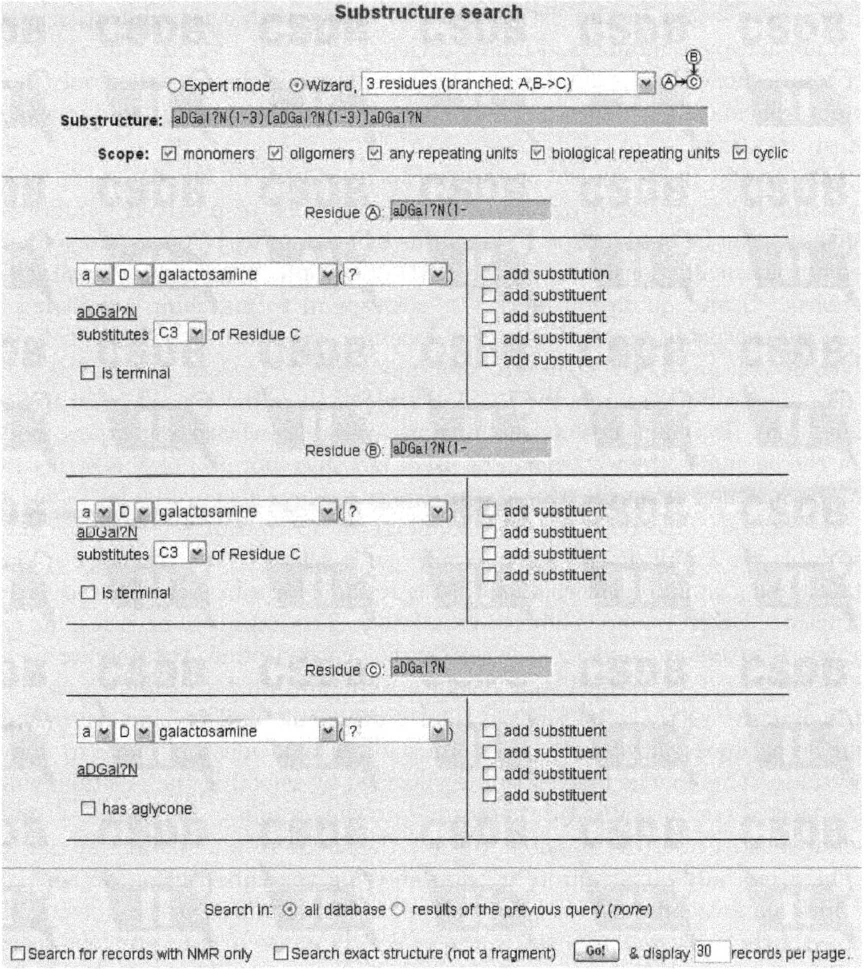

FIGURE 3.12: The substructure search interface of BCSDB. See text for details.

field. Alternatively, the `Index` button can be used, which will list all names that begin with the letters entered in the "Start with" field to its right. Text fields are also available for searching titles and keywords. Title searches can be made across abstracts by checking the "in abstracts too" checkbox. Similarly, keyword searches can be made in titles as well by checking the "in title too" checkbox. If the specific journal in which a structure was published is known, it can be selected in the list of journals, along with the publication year, volume and page numbers. Finally, if the checkbox for "Filter out records if structure elucidation is not described in the paper" is checked, then only those papers elucidating a structure (as opposed to performing additional analyses) will be returned.

For the text fields, logical operations such as AND or OR can be specified by using the ampersand (&) and bar (—) symbols, respectively. Quotation marks (") can be used for specifying phrases or names with spaces, since spaces are automatically considered as the AND operation. The wildcard characters asterisk (*) and quotation mark (?) correspond to matching any string or a single character, respectively. For example, to search for *N*- or *O*-linked glycans with any Lewis structure, one may type: `?-linked & *Lewis*`.

For substructure search, the BCSDB style of encoding is used (refer to Section 2.1.6). Two methods of searching are available: wizard and expert mode. For the beginner, the wizard can be used, but since not all possible structures can be specified using the wizard, the expert mode is also available. In wizard mode, options to choose the base topology of up to four residues are made available via a pull-down menu. When one is selected, the topology is illustrated in a graphic to the right, as in Figure 3.12 for a branched trisaccharide structure. Each residue is labeled by a letter. The details of each residue can be specified below, next to each letter. For each residue, the anomer (a for α, b for β or ? for either), the absolute configuration for residues that have optical activity (D, L, R, S or ?), the residue name and residue type, which can be pyranose, furanose, open-chain, alditol or a question mark to match anything. The linkage information is specified by selecting the position which the selected residue substitutes according to the topology. Furthermore, additional substitutions are possible to the right of the residue details by placing a check on "add substitution at" and entering the substitution details. Up to four monovalent substituents can be specified. Finally, at the bottom, the scope of the search can be set to the entire database or any previous search results, if available. Therefore, filtered searches are possible.

At the bottom of this search page, there are other links: *Make GLYDE* and *Search in GLYCOSCIENCES*. The former allows one to obtain a GLYDE 1.2 representation (see Section 2.1.9) of the entered structure, and the latter allows one to obtain search results from GLYCOSCIENCES.de.

For microorganism search, the query options for kingdom, genus, species, and strain are available. It is also possible to directly type in the strain in the text field due to the large number of strains that may be possible. Similar to substructure search, the query can be run against the entire database or any previously obtained results. A helpful list of microorganisms is also available as a link at the bottom.

3.1.5 GLYCO3D

URL http://www.cermav.cnrs.fr/glyco3d/

GLYCO3D is a resource of three-dimensional structures of glycans, polysaccharides, lectins, glycosyltransferases and glycosaminoglycan (GAG) binding proteins. Here we focus on the glycan structures. The main page for GLYCO3D provides links to the following categories of structures: monosaccharides, disaccharides, oligosaccharides, polysaccharides, lectins, glycosyltransferases and GAG binding proteins. There are other links to useful information as well, including detailed information regarding complex structures such as starch and cellulose. Clicking on any of the categories will open a new browser window for further detailed information. Since each category provides a different display, these will be described individually here. Since the polysaccharides category mainly contains polymers, these will not be described here as it is beyond the scope of this book.

3.1.5.1 Monosaccharides

The monosaccharides category consists of 18 sub-categories of major monosaccharide residues. Under each monosaccharide, further sub-categories of the available α- or β- configurations are also listed. Under these smaller sub-categories, either a listing of the corresponding structures (with modifications) or even smaller sub-categories may be listed. At the lowest level, the monosaccharide name is listed as a link. When clicked, the right-hand side will display its detailed information, such as the number of carbons, the type of cycle, configuration and any comments. Furthermore, a figure of the chair conformation of the monosaccharide as well as a Jmol viewer for the 3D structure is available along with a link to the PDB file.

3.1.5.2 Disaccharides

Disaccharides are a bit more complex compared to monosaccharides. Thus, instead of a tree-based menu of sub-categories, the user is provided with a query form of pull-down menus to search for the available disaccharides. The default structure is Fucose1-4Rhamnose. The monosaccharide names may be selected from any of the following: apiose, arabinose, fucose, galactose, glucose, mannose, rhamnose, talose and xylose. There is also the option to select ALL such that all these monosaccharides can be queried at once. When the search button is clicked, another pull-down menu containing the search results in IUPAC format is displayed.

Next, after selecting a particular structure and clicking the select button, the bottom portion of the screen displays the available low energy conformations for the selected structure. An iso-potential energy map (Ramachadran diagram) and the 3D structure using Jmol is displayed. Other values such as the total energy of each conformation and torsion angle information are also listed. A link to view the 3D structure in a wider window can also be used if the appropriate plug-in (Chime) is installed.

3.1.5.3 Oligosaccharides

This category contains information on the crystal structures of oligosaccharides. A query form similar to the disaccharides category is used for viewing the data in this category. The user may select up to two monosaccharides and the glycosidic linkage in between. The following monosaccharides are available: allose, arabinose, fructose, fucose, galactose, glucose, KDO, mannose, and xylose. There is also the option to select ALL such that any monosaccharide may match the sugar on the given side of the glycosidic linkage. An example is given in Figure 3.13 of the GlcNAcβ1-4GlcNAc structure found in the chitobiose core of *N*-glycans. The detailed information is given regarding the various notations for the structure and any references citing it. The torsion angles and the 3D structure via Jmol are also provided.

3.1.6   MonoSaccharideDB

URL http://www.dkfz.de/spec/monosaccharide-db/

MonoSaccharideDB is a comprehensive resource of monosaccharides, currently containing over 300 entries, greatly outnumbering the number of amino acids or nucleotides. Despite the small size compared to other bioinformatics resources, however, this database plays an important role in standardizing carbohydrate data formats. As was described in Section 2.1.9, the GLYDE-II standard requires the specification of residues. Although free text is an option, this would not allow for straightforward comparisons between different database formats. MonoSaccharideDB is therefore a central repository for identifying the monosaccharides used in data exchange formats.

In MonoSaccharideDB, monosaccharides consist of a basetype and a list of substituents, if applicable. The basetype corresponds to the residue size, absolute and anomeric stereochemistry and ring closure in general, but it may also contain a number of core modifications, as listed in Table 3.5.

Substituents are linked to the monosaccharide basetype by a number of linkage types, listed below:

H_AT_OH A standard O-linked substituent; the substituent replaces the hydrogen of an OH group.

DEOXY The substituent is directly linked to the basetype backbone by replacing the entire OH group.

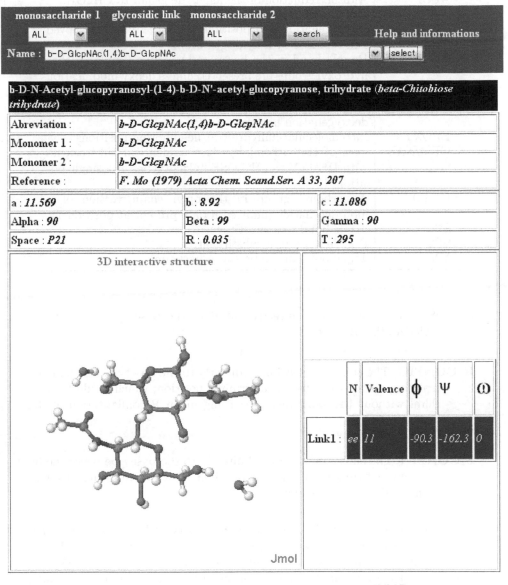

FIGURE 3.13. Display of an oligosaccharide from GLYCO3D.

TABLE 3.5: Core modifications used in MonoSaccharideDB.

Name	Description
ACID	Carboxyl (COOH) group.
ALDI	Alditol; reduction of the aldehyde group to CH2OH.
ANHYDRO	Intramolecular anhydride.
DEOXY	Deoxygenation of a position; the OH group is removed and replaced by a hydrogen atom.
EN	Double bond in the basetype backbone (implying that hydroxyl groups are preserved unless deoxy modification is explicitly stated).
ENX	Double bond in the basetype backbone with unknown deoxygenation pattern.
EPOXY	Intramolecular anhydride at neighboring positions.
GEMINAL	Loss of stereochemistry due to identical substituents with DEOXY and H_LOSE linkage types at a single position.
KETO	A carbonyl group in the open chain version of a monosaccharide (omitted if only present at position 1).
SP	Triple bond to a substituent.
SP2	Double bond to a substituent.
YN	Triple bond in the basetype backbone.

H_LOSE The substituent is directly linked to the basetype backbone by replacing the hydrogen atom.

R_CONFIG The substituent is linked directly to the basetype backbone by replacing a hydrogen atom at a terminal position, which would be non-chiral without the substituent, resulting in an R-configuration of the carbon.

S_CONFIG The substituent is linked directly to the basetype backbone by replacing a hydrogen atom at a terminal position, resulting in an S-configuration of the carbon.

Several options are available for querying MonoSaccharideDB. Monosaccharides, substituents, and atomic elements are the main options. Under monosaccharides, options to query by name/id, properties and using a monosaccharide builder are available. Names can be entered in a variety of formats, including CarbBank, GLYCOSCIENCES.de, MonoSaccharideDB, GlycoCT, CFG, BCSDB and PDB. Substitutions may be optionally specified in the name query. The monosaccharide builder is a form where after the selection of the ring size, a variety of options for building the backbone, the ring closure and substituents are provided, as in Figure 3.14.

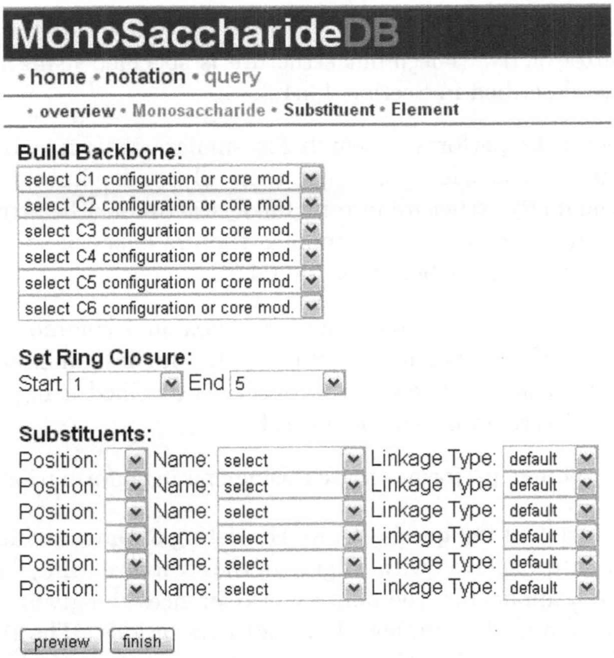

FIGURE 3.14: The MonoSaccharide Builder of MonoSaccharideDB for performing queries.

3.1.7 GlycomeDB

URL http://www.glycome-db.org/

GlycomeDB is the latest of the glycan structure databases that are publicly available (Ranzinger et al. (2008)). It consists of all of the major glycan structure databases including KEGG GLYCAN, CFG, GLYCOSCIENCES.de, and BCSDB, and it provides structure and species information stored in an integrated database. The GlycomeDB system is continually updated automatically by retrieving the latest information from all databases and performing semi-automatic integration of the latest data.

The main page to GlycomeDB provides links to access and query the database and to download the data in a variety of formats. The "Database" link displays a multitude of search options for querying GlycomeDB, described below:

Search by database ID: perform a query using known glycan structure IDs from external databases included in GlycomeDB.

Exact structure search: perform a search for structures matching the input structure exactly. The input structure is specified using a Java-based interface described in Section 4.5.1.3.

Substructure search: perform a search for structures containing the input structure exactly. The input structure is specified using a Java-based interface described in Section 4.5.1.3.

Similarity search: perform a search for similar structures by comparing links (monosaccharide pairs and their glycosidic linkage). Breaking down the query structure into its links, search for other structures containing the same links. The input structure is specified using a Java-based interface described in Section 4.5.1.3.

MCS search: perform a search using the maximum common substructure algorithm. Find structures containing the highest maximum common substructure score. The input structure is specified using a Java-based interface described in Section 4.5.1.3.

Search by species: perform a search using species information.

The result of any of the queries above is a listing of the matching structures from the database displayed with a figure of the structure using CFG notation, a match score if applicable, the number of references the given structure has in all databases, and the number of species annotations. The results of this query may be refined using the Complex Query System (CQS). There are three options in this system: (1) retrieve the intersection of the results with another query, (2) retrieve the union of the results with another query, or (3) retrieve the complement of the results. For (1) and (2), a second search may be performed using one of the following: (a) search by database, (b) search by species, (c) search by substructure, (d) search by maximum common substructure, or (e) search by similar substructure. For (3), the entries that were NOT retrieved by the current query are obtained.

Clicking on the detailed information for a particular structure entry, several options become available. First, the image of the structure may be changed to any of the following styles: CFG, Oxford, IUPAC or GlycoCT. These may be displayed in either PNG or SVG formats. Second, the species annotations are displayed with links to the taxonomy information at NCBI. Third, links to other databases corresponding to the selected structure are listed. Fourth, the structure can be obtained in other ASCII formats such as GlycoCT{condensed} (Section 2.1.8), GlycoCT{xml} or GLYDE-II XML formats (Section 2.1.9).

3.2   Glyco-gene databases

Glycans are often recognized by proteins and other pathogens, thus signaling various events in the biological system. There are a number of databases containing information regarding glycan-related proteins, and these will be described in this section.

3.2.1   KEGG BRITE

URL http://www.genome.jp/kegg/brite.html

KEGG contains information regarding glycan binding proteins classified as a functional hierarchy in the KEGG BRITE database, which contains classifications of all known proteins based on functional information from pathways and sequence similarity. Glycan-related hierarchies include glycosyltransferases, glycosyltransferase/glycosidase reactions, and glycan-binding proteins. Note that glycosyltransferases and glycan-binding proteins are hierarchies of genes whereas glycosyltransferase/glycosidase reactions is one of reactions. Thus the protein-based hierarchies contain the list of orthologous gene groups (KOs) contained in each hierarchy. Under each KO group, the individual genes are provided along with their annotations.

In contrast, the reactions hierarchy are organized such that groups of similar reactions are clustered into the same groups. For example, reactions involving the transfer of glucose and fucose are classified under "Glucosyltransferase reactions" and "Fucosyltransferase reactions," respectively.

3.2.2   CFG

As mentioned earlier in Section 3.1.3, the CFG provides information on glycan binding proteins (GBPs) and glycosyltransferases under the *CFG Databases* area. The GBP data provide all known data regarding lectins including C-type lectins, galectins and siglecs, to name a few. When a particular class of lectins is clicked, the list of GBP molecules and their corresponding names and species are listed, as in Figure 3.15 for Collectins.

In this list, by clicking on the name of a specific protein, detailed information can be retrieved, organized into the following tabs: General, Reference, Genome, Proteome, Glycome, and Biology. The General information tab provides information for specific proteins including the category and sub-family of the protein, other names for the protein, the species in which it is found, and a concise summary annotating the function, provided by experts (usually participating investigators of the consortium). The Reference tab provides links to other databases for the same protein as well as a link to the Entrez Gene and OMIM entries from the National Center for Biotechnology Information (NCBI) to search for related literature using the protein's known names. Genomic information is provided, including Gene Ontology (GO) terms, nucleotide accession numbers, chromosome number, etc., as well as links to search for genes in expression databases and BLAST. The Proteome data include the amino acid sequence, PDB links to 3D structures of the protein and molecular weight. The Glycome information includes explanations

Glycan Binding Proteins

3-Collectin Other GBP Families

Results 1 - 15 of 15

Results Per Page: 25 [v] [View]   Sort results by: View GBP Molecule [v] [View]

Filter Results where [View GBP Molecule [v]] contains [   ] [View]

View GBP Molecule	CFG Name	Species	List Date
cbp_hum_Ctlect_225	Collectin K1	Human	05/21/2004
cbp_hum_Ctlect_226	Collectin L1	Human	05/21/2004
cbp_hum_Ctlect_227	Mannose-binding protein	Human	05/21/2004
cbp_hum_Ctlect_228	Pulmonary surfactant protein SP-D	Human	05/21/2004
cbp_hum_Ctlect_232	Pulmonary surfactant protein SP-A1	Human	05/21/2004
cbp_mou_Ctlect_168	Mannose-binding protein A	Mouse	05/21/2004
cbp_mou_Ctlect_169	Mannose-binding protein C	Mouse	05/21/2004
cbp_mou_Ctlect_170	Pulmonary surfactant protein SP-D	Mouse	05/21/2004
cbp_mou_Ctlect_352	Collectin L1	Mouse	10/17/2004
cbp_mou_Ctlect_353	Collectin K1	Mouse	10/17/2004
cbp_mou_Ctlect_355	Pulmonary surfactant protein SP-A	Mouse	10/17/2004
cbp_rat_Ctlect_00135	Mannose-binding protein A	Rat	01/06/2004
cbp_rat_Ctlect_773	SLex-Mannose-Binding Protein (SLex-MBP-2)	Rat	09/08/2006
cbp_rat_Ctlect_774	SLex-Mannose-Binding Protein (SLex-MBP-3)	Rat	09/08/2006
cbp_rat_Ctlect_775	SLex-Mannose-Binding Protein (SLex-MBP-4)	Rat	09/08/2006

FIGURE 3.15: The list of collectins provided by the CFG. Each molecule is listed with its corresponding names and species information.

of glycan binding specificity, possible counter receptors, links to any glycan-GBP interaction data provided by the CFG as well as PDB links for ligands. Finally, the Biology tab provides any other expert knowledge regarding the biology of the protein, such as pathology.

The glycosyltransferase data in CFG is provided as a graphic of a glycan structure containing all possible linkages, as in Figure 3.16. All available enzyme information is categorized by the glycan type on the left. When one is clicked, all possible linkages are displayed in the image on the right. Clicking on a monosaccharide in the image will display the species-specific information for the enzyme that is known to synthesize the linkage. When a particular enzyme is clicked, a page similar to the glycan-binding molecule page will be displayed with tabs for General, Reference, Genome, Proteome and Activity. The General tab contains the EC number and any given names for the enzyme. The Reference tab provides links to other databases including PubMed, EXPASy, KEGG, CAZy, NCBI's Entrez Gene and SwissProt. The Genome tab provides genomic information such as links to other genomic databases, the nucleotide sequence and a link to BLAST the DNA sequence. Likewise, the Proteome tab provides the amino acid sequence and a link to BLAST it, in addition to a link to the corresponding Swiss-Prot entry. Finally, the Activity tab provides known information provided by experts regarding this enzyme and a link to the KEGG Enzyme entry.

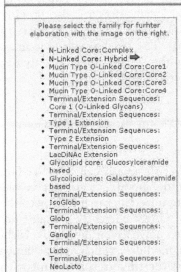
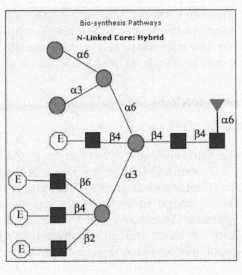

FIGURE 3.16: The graphical interface for the CFG Glyco Enzyme data. All available enzyme information is categorized by the glycan type on the left. When clicked, all possible linkages are displayed in a graphic on the right.

3.2.3   GGDB

URL http://riodb.ibase.aist.go.jp/rcmg/ggdb/

The GlycoGene DataBase (Narimatsu (2004)) is developed by the Glyco-Gene Function Team of the Research Center for Medical Glycoscience at the National Institute of Advanced Industrial Science and Technology (AIST) in Japan. It contains information regarding genes associated with glycan synthesis, which includes glycosyltransferases, sugar-nucleotide synthases, sugar-nucleotide transporters, etc. All genes have been identified, cloned and characterized from human samples, totaling almost 200 genes. From the top page, the listing of all available genes are presented in hierarchical form at the left. Clicking on a gene will display a list of homologous genes with links to NCBI's Protein and RNA data, EC number and CAZy ID. If available, citations on the acceptor substrates, an applet visualizing the enzymatic reaction, and information regarding the expression such as localization are also provided. An example is the α2,3-sialyltransferase ST3GAL1 in Figure 3.17, where two different substrate specificities are illustrated. The red linkages fade in and out to illustrate where this enzyme transfers the sugar. Enzyme activity as measured by the developers is also listed where available.

3.2.4   CAZy

URL http://www.cazy.org/

The Carbohydrate-Active Enzymes (CAZy) database contains the families of structurally-related enzymes that degrade, modify, or create glycosidic bonds (Cantarel et al. (2009)). The data is organized hierarchically according to the structural similarity of the enzymes. CAZy currently contains data on glycoside hydrolases, glycosyltransferases, polysaccharide lyases, carbohydrate esterases, and carbohydrate-binding modules, which are all derived from publicly available genome sequences. At the time of this writing, there are 91 families of glycosyltransferases (GT families), 114 families of glycoside hydrolases (GH), 21 families of polysaccharide lyases (PL), 16 families of carbohydrate esterases (CE), and 53 families of carbohydrate binding molecules (CBM).

For each family, its known activities, 3D structure, and statistics in terms of number of entries in other databases and taxonomic classifications are provided. For each enzyme, the EC number (if available), the organism, and any IDs to other databases are also listed. A figure of Glycosyltransferase Family 1 (GT1) is given in Figure 3.18.

Family	Sialyltransferases
Official Symbol	ST3GAL1
Alias Symbol	ST3O, SIAT4A, SIATFL, MGC9183, ST3GallA, Gal-NAc6S, ST3GalA.1
Designation	CMP-NeuAc:beta-galactoside alpha-2,3-sialyltransferase I
Organism	H.sapiens

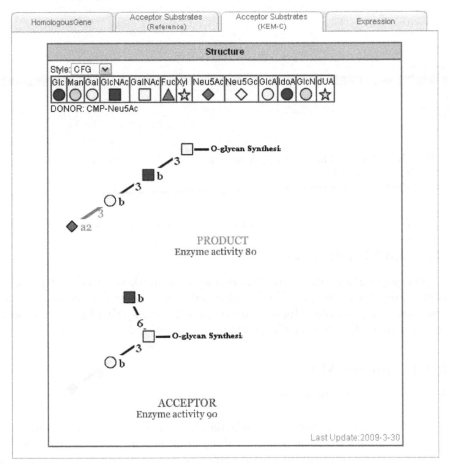

FIGURE 3.17: The substrate specificity of ST3GAL1 as presented by CGDB.

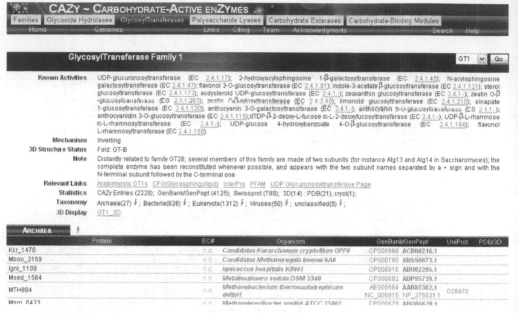

FIGURE 3.18:   The list of enzymes classified into glycosyltransferase family 1 (GT1) in CAZy.

3.3   Lipid databases

Just as glycans on glycoproteins often serve as markers for cellular recognition, so do glycans on glycolipids. Although few compared to proteins, lipid databases also contain information on carbohydrates (glycolipids), so these will be briefly discussed in this section.

3.3.1   SphingoMAP©

URL http://www.sphingomap.org/

SphinGOMAP© is not so much a database as it is a resource that provides an evolving pathway map for sphingolipid biosynthesis (Merril (2005)), including glycosphingolipids. This map contains approximately 450 glycan structures arranged such that the pathway for each sphingolipid sub-type is distinguished (see Section 1.3 on the different glycan sub-classes). It uses the symbol annotation suggested by the CFG to represent glycans, and it can be downloaded in a number of formats, including PDF and JPG. One should note that, as is the case for most pathway data, since this map is based on the literature, it is subject to change based on new findings and user feedback.

3.3.2 LipidBank

URL http://lipidbank.jp/

The LipidBank database is a collection of lipids originally collected and organized by the Japanese Conference on the Biochemistry of Lipids (JCBL) (Watanabe et al. (2000)). It currently includes 16 categories of over 7000 molecules. Glycan-related lipids include almost 700 glycolipids and over 700 lipopolysaccharides. All the data can be browsed from the main web page, and searches can be performed once a specific category (or the "All data" category) is clicked (see Figure 3.19).

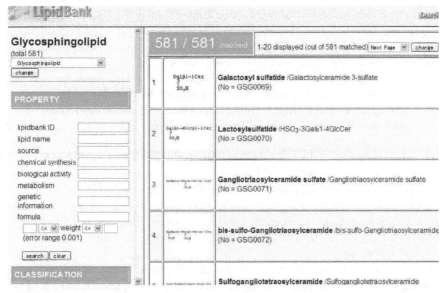

FIGURE 3.19: The list of glycosphingolipids as categorized by LipidBank.

To perform a search in the glycosphingolipid category, for example, a keyword text search for any of the following properties can be performed: lipidbank ID, lipid name, biological source, chemical synthesis, biological activity, metabolism, genetic information, chemical formula, and a range of molecular weights. Selections can also be filtered by choosing a classification: neutral, acidic, alkaline or amphoteric. Glycan sequences can also be selected from a long list of candidates under "Sugar chain series," or a particular glycan sequence can be specified under "Other" where a text field is available with options to indicate the reducing or non-reducing ends to match the structure. Finally, a search by composition can also be run from the "Number" section.

3.3.3   LMSD

URL http://www.lipidmaps.org/data/structure/

LMSD, which stands for LIPID MAPS structure database (Sud et al. (2006)), was developed by the LIPID MAPS Consortium, which was given charge of characterizing the lipid section of the metabolome by developing an integrated metabolomic system that was capable of characterizing global changes in lipid metabolites. As a result, LMSD currently contains over 10,000 structures of lipids from four sources: the LIPID MAP's Consortium, those identified by LIPID MAPS experiments, computationally generated structures, and biologically relevant lipids manually curated from LipidBank and other public resources. The LIPID MAPS consortium members developed a classification system for lipids consisting of eight categories: (i) fatty acyls, (ii) glycerolipids, (ii) glycerophospholipids, (iv) sphingolipids, (v) sterol lipids, (vi) prenol lipids, (vii) saccharolipids and (viii) polyketides. From the main page, LMSD data can be browsed, searched, and even downloaded.

Structures may be browsed by classification or searched based on text keywords, ontology, or by structure. The text-based search takes as input any of the following properties: Lipid Map (LM) ID, common or systematic name, a range of masses, and formula. Pull-down menus can also be used to select a particular category, from which further classifications of main class and subclass can be selected. The sphingolipids category is divided into 10 main classes, including sphingoid bases, ceramides, phosphosphingolipids and various glycosphingolipids, as shown in Figure 3.20.

The ontology search is similar to the text-based search, except that there are additional options to specify various limits to the number of chemical substituents, such as the number or range of carbons or rings, the number or range of esters, etc.

Finally, structure search requires the use of one of the available structure drawing tools: MarvinSketch Applet, JME Applet or ChemDrawPro. These tools require either Java applet support to be enabled in the web browser or a plugin to be installed (in the case of using ChemDrawPro). Structures can be searched using exact match or substructure search, and users may also specify an LM ID or name to include in the search. From any search result, as in Figure 3.21, clicking on the LM ID will display the entry page where an image of the structure is drawn along with its particular properties, including links to PubChem and KEGG (Figure 3.22).

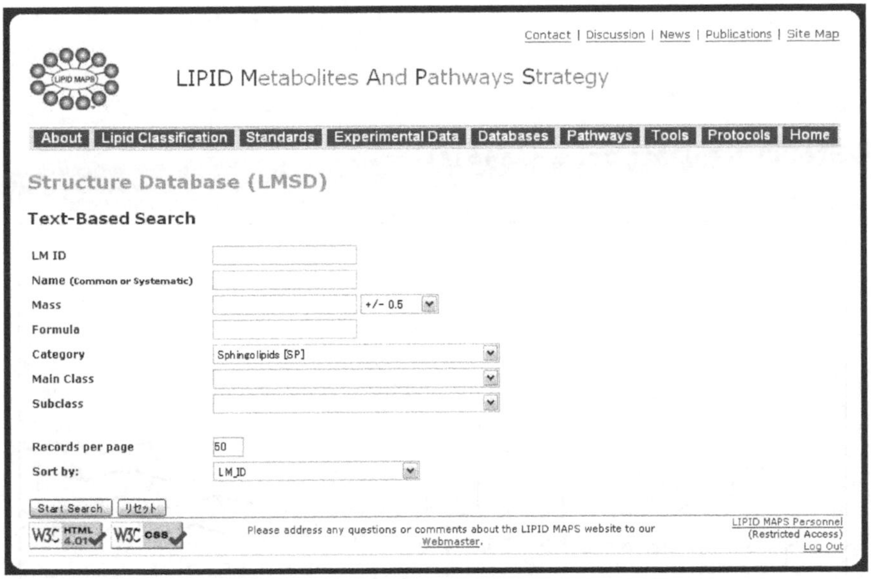

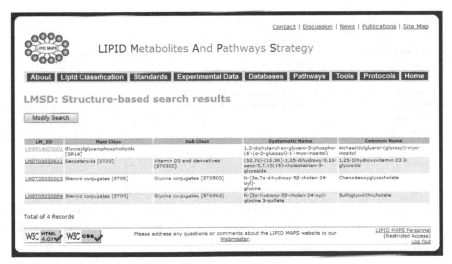

FIGURE 3.20: The LMSD text-search interface when the sphingolipid category is selected. A listing of ten main classes becomes available.

FIGURE 3.21: The LMSD results page of a search. Each entry is listed with its ID linked to its entry page, class and subclass information, and names for the structure.

Structure Database (LMSD)

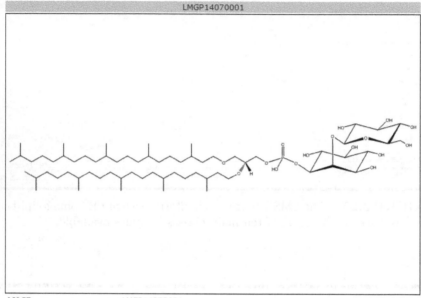

LMGP14070001

LM ID	LMGP14070001
Common Name	Archaetidylglyerol-(glycosyl)-myo-inositol
Systematic Name	1,2-diphytanyl-sn-glycero-3-phospho-(6'-(α-D-glucosyl)-1'-myo-inositol)
Exact Mass	1196.90
Formula	$C_{65}H_{129}O_{16}P$
Category	Glycerophospholipids [GP]
Main Class	Glycosylglycerophospholipids [GP14]
Sub Class	
Synonyms	-
PubChem Substance ID (SID)	-
Status	Active
KEGG ID	-

View Structure using MarvinView Applet   JmolApplet   ChemDraw   (?)

FIGURE 3.22:   The LMSD entry page for a glycerophospholipid, where the image of the structure is drawn, followed by detailed information regarding the structure.

3.4   Lectin databases

Lectins were introduced in Section 2.2. The currently available databases of lectins are briefly introduced in this section.

3.4.1   Lectines

URL http://www.cermav.cnrs.fr/lectines/

The Lectines database (Krengel and Imberty (2007)) is part of GLYCO3D, which was introduced in Section 3.1.5. The 3D structures of lectins in Lectines are organized hierarchically, with the major categories including algae, animal, bacteria, fungi and yeast, plants, and viruses. For a particular lectin, an explanation of any 3D structures found in PDB is provided. Detailed information includes the PDB code, species, resolution and 3D images of the selected lectins, along with links to PDB and other related databases. Data may only be browsed; there is no search or download function for this resource.

3.4.2   Animal Lectin DB

URL http://www.imperial.ac.uk/research/animallectins/

The Animal Lectin DB is a genomics resource, developed by Dr. Kurt Drickamer of Imperial College London, for animal lectins, consisting of two parts: (1) structures and functions of animal lectins, and (2) C-type lectin-like domains. Part 1 provides a general introduction to lectin families including their location, binding specificity and evolution. Information for specific lectin families that are often not found in other databases are also found here. Part 2 provides detailed information regarding the C-type lectin-like domain (CTLD), including CTLD evolution, identification and classification in mammals.

3.5   Others

Carbohydrate antigens/antibodies and organism-specific carbohydrates have also been compiled in some databases. These carbohydrate-related databases are presented in this section.

3.5.1   GlycoEpitopeDB

URL http://www.glyco.is.ritsumei.ac.jp/epitope/

The GlycoEpitope database is collection of all known carbohydrate anti-gens, or glyco-epitopes, and antibodies (Kawasaki et al. (2006)). Epitope search includes text search for the following tags: epitope ID, composition, biological source, disease and reference information. As an example, a search for Lewis structures in the Epitope field will produce 21 entries for the differ-ent Lewis epitopes, as displayed in Figure 3.23. Clicking on an epitope ID, such as EP0009 for 3'-Sulfo Lewis[a], a tab-based window of all information known for this epitope will open, as in Figure 3.24. For each entry, detailed information is available regarding the epitope's localization, related enzymes, antibodies and glycoconjugates such as glycoproteins and glycolipids, in addi-tion to references. In the example, the structure is provided as an image and general information such as the carbohydrate sequence, aliases, history, molec-ular weight, composition, species, tissue and cellular localization, cell lines, receptor, function, disease and application information are listed. Under the Antibody tab, the corresponding known antibodies and their information are displayed. From the link for the antibody name, detailed antibody informa-tion will be displayed, as described later. The Glycoprotein tab lists all known protein carriers of this epitope, with links to NCBI Gene and Swissprot. The Glycolipid tab likewise lists all known lipid carriers. Under the Enzyme tab, both biosynthetic and degradation enzymes known for this epitope are dis-played with links to EC as well as NCBI Gene and Swissprot. Finally the References tab lists the known literature describing this epitope.

Antibodies may also be searched, which takes as input text values for anti-body ID or name, or a pull-down menu containing the registered antibodies may be used to select an antibody. Figure 3.25 is an example of an antibody entry page, where the corresponding epitope information is briefly listed along with information regarding the antibody such as the species in which it can be found, the category (e.g., Monoclonal), isotypes, and references. There is also a link to suppliers and information regarding known applications of the given antibody.

3.5.2   ECODAB

URL http://www.casper.organ.su.se/ECODAB/

The ECODAB database contains structures of the repeating units compris-ing the *O*-antigen which makes up a part of the lipopolysaccharides (LPS) found on the outer membrane of *E. coli*. This database was created in order to establish the uniqueness of the structures of *O*-antigenic polysaccharides. For each glycan, information on structures, NMR data and cross-reactivity relationships are provided. A simple search interface is also available, where any terms can be entered, separated by a space. Values are matched to the closest numerical value in the record, while strings are matched as substrings (e.g., glc matches both glucose and N-acetylglucosamine). Results are listed with the structure and citations, as in Figure 3.26.

FIGURE 3.23: The epitope search results for Lewis structures from Gly-coEpitope DB consists of 21 epitopes.

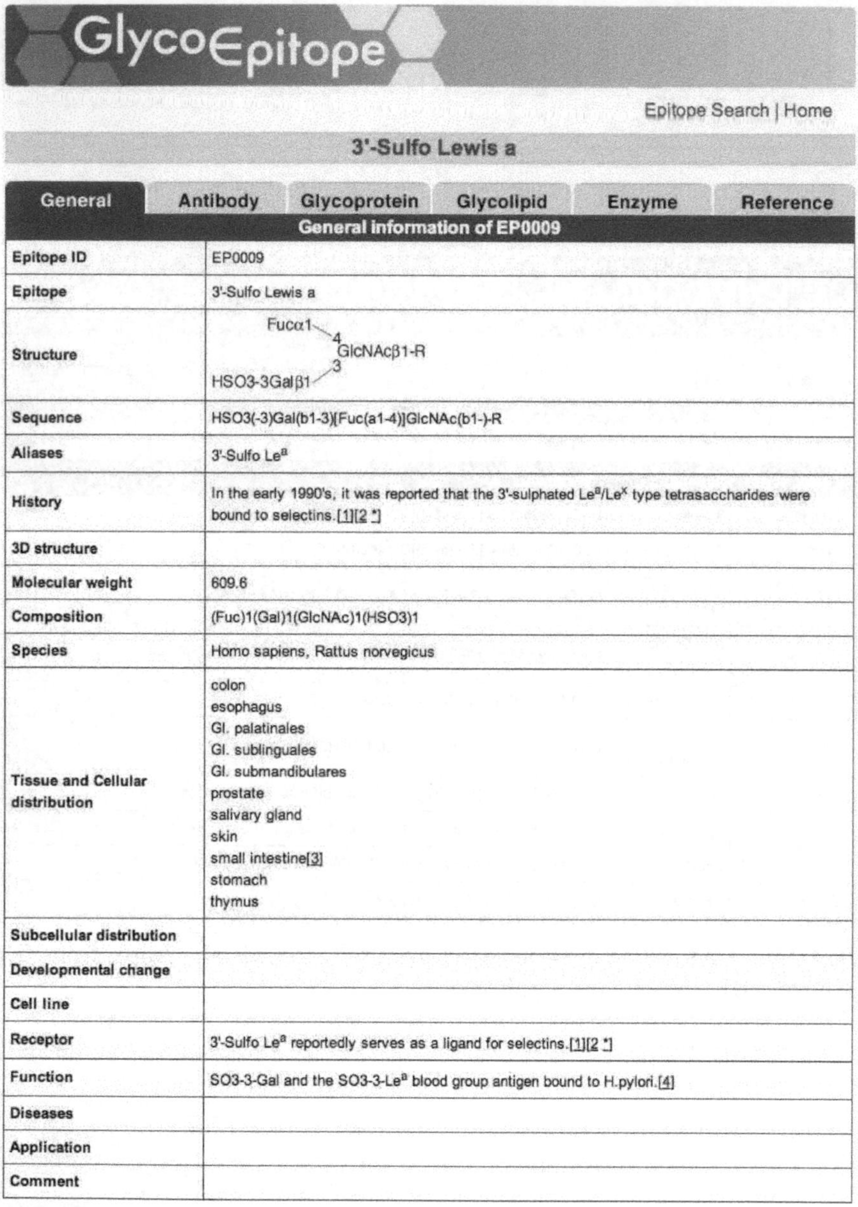

GlycoEpitope

3'-Sulfo Lewis a

General	Antibody	Glycoprotein	Glycolipid	Enzyme	Reference

General information of EP0009

Epitope ID	EP0009
Epitope	3'-Sulfo Lewis a
Structure	Fucα1 4 GlcNAcβ1-R 3 HSO3-3Galβ1
Sequence	HSO3(-3)Gal(b1-3)[Fuc(a1-4)]GlcNAc(b1-)-R
Aliases	3'-Sulfo Le^a
History	In the early 1990's, it was reported that the 3'-sulphated Le^a/Le^x type tetrasaccharides were bound to selectins.[1][2 *]
3D structure	
Molecular weight	609.6
Composition	(Fuc)1(Gal)1(GlcNAc)1(HSO3)1
Species	Homo sapiens, Rattus norvegicus
Tissue and Cellular distribution	colon esophagus Gl. palatinales Gl. sublinguales Gl. submandibulares prostate salivary gland skin small intestine[3] stomach thymus
Subcellular distribution	
Developmental change	
Cell line	
Receptor	3'-Sulfo Le^a reportedly serves as a ligand for selectins.[1][2 *]
Function	SO3-3-Gal and the SO3-3-Le^a blood group antigen bound to H.pylori.[4]
Diseases	
Application	
Comment	

FIGURE 3.24: The epitope entry page for 3'-sulfo Lewis^a.

F2	

Antibody ID		AN0170
Epitope	Epitope ID	EP0009
	Epitope Name	3'-Sulfo Lewis a
	Recognition region	HSO3(-3)Gal(b1-3)[Fuc(a1-4)]GlcNAc-R
	Immunoprecipitation	
	Immunoblot	[1]
	ELISA/RIA	
	Flow cytometry	
	Histochemistry	[2][3]
	Comment	
Species		Mouse
Category		Monoclonal
Isotype		IgM
Availability		MONOSAN abcam Funakoshi -Japanese
Application		

1	Veerman EC, Valentijn-Benz M, van den Keybus PA, Rathman WM, Sheehan JK, Nieuw Amerongen AV. Immunochemical analysis of high molecular-weight human salivary mucins (MG1) using monoclonal antibodies. Arch Oral Biol. 1991 ;36(12):923-32 PMID: 1722666
2	Veerman EC, Bolscher JG, Appelmelk BJ, Bloemena E, van den Berg TK, Nieuw Amerongen AV. A monoclonal antibody directed against high M(r) salivary mucins recognizes the SO3-3Gal beta 1-3GlcNAc moiety of sulfo-Lewis(a): a histochemical survey of human and rat tissue. Glycobiology. 1997 Feb ;7(1):37-43 PMID: 9061363
3	Atlas P, Deutsch V, Lieberman Y, Neufeld HN. False aneurysm of the left atrium after closed mitral commissurotomy: diagnosis by cineangiocardiogrpahy. Report of one case treated surgically. J Cardiovasc Surg (Torino). 1976 Mar ;17(2):170-3 PMID: 1262387

FIGURE 3.25: The antibody entry page for F2, an antibody to 3'-Sulfo Lewis^a.

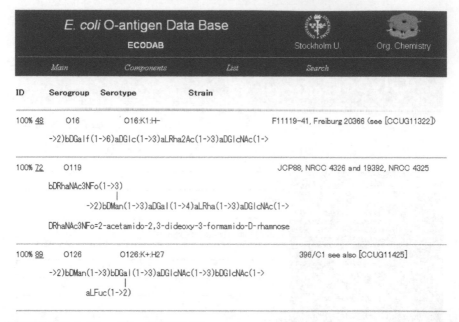

FIGURE 3.26: The ECODAB search results page, with matched entries listed alongside the citations corresponding to the entry.

3.5.3   SugarBindDB

URL http://sugarbinddb.mitre.org/

SugarBindDB is a pathogen sugar-binding database containing carbohydrate structures that pathogenic organisms are known to bind. To search the database, a variety of search options are available. First, the type of data that should be returned can be specified, including pathogen/toxin, lectin/adhesin, carbohydrate/ligand, publication year and citation. In the search parameters, one or more pathogens or toxins may be selected in the list provided. Specific items may be searched for using the "Find in List" search text field. Furthermore, up to four residues of a carbohydrate sequence may be specified using the pull-down menus under the Pathogen/Toxin list. Finally, author and publication year(s) may be entered at the bottom. The results of a search will display the items selected, and links to citations are provided.

Chapter 4

Glycome Informatics

This chapter is the main focus of this book, introducing the informatic methods that have been developed over the years to analyze glycans. In Section 4.1, the terminology and notations used in this chapter will be presented. In Section 4.2, computer theoretic algorithms applied to glycan structure comparison will be introduced. This is in contrast to Section 4.3, where bioinformatic methods utilizing genomic data are described. Section 4.4 introduces data mining methods for glycan analysis, and the last section lists the tools that have been developed for glycome informatics analysis.

4.1 Terminology and notations

In order to prepare readers for this chapter on algorithmic techniques, some definitions of terminology and notations will be introduced here. These will be used throughout this chapter. First, a *tree*, say T, is defined as the set of nodes V connected by edges E such that no cyles exist. Thus T can be described by (V, E). For glycans, nodes correspond to monosaccharides, and edges correspond to glycosidic linkages. In rare cases, cyclic glycans do exist, but these will not be considered here.

A tree is *rooted* if there exists a particular node (the *root node*) in the tree from which all other nodes emanate. In the case of glycans, the monosaccharide at the reducing end would be the root node. The nodes at the opposite end of the tree are called *leaves*. These would correspond to the sugars at the non-reducing end of glycans. In analogy to phylogenetic trees, then, the root is considered an *ancestor*, and all nodes along the tree towards the leaves are *descendants* of the root. The immediate descendant m of any node n is considered the *child* of n, and n is considered the *parent* of m. Subsequently, the child of m would be considered the *grandchild* of n. All the children of the same node are considered *siblings* of one another. If the siblings have a defined order, the tree is considered an *ordered tree*. Furthermore, if labels are attached to each node (and perhaps each edge), then the tree is a *labeled tree*. In the case of glycans, nodes are labeled with the monosaccharide name, and edges with the glycosidic bond information. Alternatively, edges may be unlabeled when nodes are labeled with both the monosaccharide name and the information regarding glycosidic linkage with the parent. Choosing any node n in a tree t and extracting all descendants of n, the resulting set of nodes and edges form a *subtree* of t rooted at n. Note that glycans are considered labeled, ordered, rooted trees.[1]

4.2 Algorithmic techniques

One of the first computer theoretic algorithms applied to glycan structure analysis was a dynamic programming approach for aligning tree structures. For beginners, an introduction to sequence alignment using dynamic programming is provided in Appendix A.1.

4.2.1 Tree structure alignment

One of the first applications of tree-structure alignment to glycans was KCaM, or KEGG Carbohydrate Matcher (Aoki et al. (2003, 2004)). This program developed out of the need for a search tool similar to BLAST for the KEGG GLYCAN database (see Section 3.1.1), since sequence alignment algorithms could not be directly applied to glycan structures. KCaM takes a variety of parameters, depending on the type of search that the user may want to perform. The two major parameters are for Exact or Approximate match, and Global or Local search. In either case, the basic idea is that two nodes u and v of two trees could be compared based on the mapping of the respective children.

For the local exact matching algorithm, the concept of the maximum common subtree (MCST) is used to find the largest common subtree between the query and candidate glycan structures. The following is the dynamic programming algorithm for finding the MCST between two trees T_u and T_v.

$$R[u, 0] = 0,$$
$$R[0, v] = 0,$$

$$R[u, v] = 1 + \max_{\psi \in \mathcal{M}(u,v)} \left\{ \sum_{u_i \in C(u)} R[u_i, \psi(u_i)] \right\}.$$

---

[1]Note that the ordering of siblings in glycans may be defined differently based on the intent of the computation. For example, structural comparisons may make use of the carbon numbers on the parent to which a child is attached, whereas pathway analyses may prefer to use the order by which children are transferred to the parent by glycosyltransferases.

Here, $\mathcal{M}(u,v)$ is the set of all possible mappings[2] between the children of u and v, $C(x)$ is the set of children of node x, and $w(u,v)$ is the similarity score between u and v. This similarity score can be defined by a weighting matrix (see Section 4.2.2) between parent-child monosaccharide pairs and their glycosidic linkages. The simplest weighting matrix would produce scores of zero (0) for a mismatch and one (1) for a match. Note that the root node has a null parent.

All possible combinations of children can be compared using $\mathcal{M}(u,v)$ and the mapping that produces the maximum score would be chosen in the dynamic programming calculation. As a result of traversing along all combinations of nodes in the two given trees, the score for the maximum common subtree can be found from the maximum value of $R[u_i, v_j]$ over all i and j. Furthermore, the actual alignment can be obtained by backtracking along the values in the matrix that contributed to the maximum score.

The global exact matching algorithm is a recursive call to the local exact matching algorithm using the unmatched portions of the given trees. Thus both matches in the core structure as well as the non-reducing end of glycans can be found exactly.

In contrast to the MCST algorithm, the dynamic programming algorithm for local approximate matching allows gaps in the alignment, as described by the following.

$$Q[u,0] = 0,$$
$$Q[0,v] = 0,$$

$$Q[u,v] = \max \begin{cases} 0, \\ \max_{v_i \in C(v)} \{Q[u, v_i] + d(v)\}, \\ \max_{u_i \in C(u)} \{Q[u_i, v] + d(u)\}, \\ w(u,v) + \max_{\psi \in \mathcal{M}(u,v)} \left\{ \displaystyle\sum_{u_i \in C(u)} Q[u_i, \psi(u_i)] \right\}. \end{cases}$$

Here, the cost of deleting a node v, $d(v)$, corresponds to the cost of v regardless of the subtree rooted at that node. The value of $w(u,v)$ represents the similarity between nodes u and v. Since glycosidic linkage information must also be considered, linkage information is compared in computing $w(u,v)$ as in Equation 4.1. In this equation, for glycosidic bond "α1-4" for example, *cCarbon* corresponds to α1 and *pCarbon* corresponds to 4, the hydroxyl group

---

[2] A mapping is defined as the combination of pairs between two sets of objects. In the case of glycans, the children of node u and those of node v are mapped to one another in a variety of combinations. As an example, let us suppose that the children of $u = \{u_1, u_2\}$ and the children of $v = \{v_1, v_2, v_3\}$. Then the following sets of possible mappings make up $\mathcal{M}(u,v)$: $\{(u_1, v_1), (u_2, v_2)\}$, $\{(u_1, v_1), (u_2, v_3)\}$, $\{(u_1, v_2), (u_2, v_1)\}$, $\{(u_1, v_2), (u_2, v_3)\}$, $\{(u_1, v_3), (u_2, v_1)\}$, and $\{(u_1, v_3), (u_2, v_2)\}$.

on the parent to which the child is linked. The values for the parameters α and β are set according to the extent that matches should be weighted.

$$w(u,v) = \max \begin{cases} 0, \\ \alpha \cdot \delta(name(u), name(v)) \\ \quad - \beta \quad (1 - \delta(cCarbon(p(u), u), cCarbon(p(v), v))) \\ \quad - \beta \cdot (1 - \delta(pCarbon(p(u), u), pCarbon(p(v), v))) \end{cases} \tag{4.1}$$

In contrast, the algorithm for global approximate matching penalizes the deletion of a node by taking the cost of deleting that node and all unmatched portions of the subtree rooted at that node. Thus the algorithm for global approximate alignment of trees T_1 and T_2 becomes the following:

$$Q[u,0] = \sum_{u_i \in T_1(u)} d(u_i),$$

$$Q[0,v] = \sum_{v_i \in T_2(v)} d(v_i),$$

$$Q[u,v] = \max \begin{cases} \max_{v_i \in C(v)} \left\{ Q[u,v_i] + d(v) + \sum_{v_j \in C(v)-\{v_i\}} Q[0,v_j] \right\}, \\ \max_{u_i \in C(u)} \left\{ Q[u_i,v] + d(u) + \sum_{u_j \in C(u)-\{u_i\}} Q[u_j,0] \right\}, \\ w(u,v) + \max_{\psi \in \mathcal{M}(u,v)} \left\{ \sum_{u_i \in C(u)} Q[u_i, \psi(u_i)] + \sum_{v_i \in C(v)-\psi(C(u))} Q[0,v_i] \right\}. \end{cases}$$

The resulting score is $\max_{u,v} Q[u,v]$, and the alignment can be backtracked by finding those nodes that contributed to the resulting score.

4.2.2 Linkage analysis using score matrices

With the development of an algorithm for assessing the similarity of glycan tree structures, the analysis of glycosidic linkages became possible. In particular, the similarity of monosaccharides and their linkages, corresponding to amino acid similarity as represented by amino acid substitution matrices such as PAM (Dayhoff et al. (1983)) and BLOSUM (Henikoff and Henikoff (1992)), was considered. However, monosaccharides alone do not make up glycan structures; the glycosidic linkage conformation information should also be taken into consideration. Therefore, an appropriate glycan score matrix would be one where such linkages and the monosaccharides being linked should be used as the basic unit for comparison.

In generating the BLOSUM score matrix, a database of protein families from the BLOCKS database was used as the base set of amino acid sequences from which to compute amino acid similarities. The interested reader may refer to Appendix A.2 for an introduction to the BLOSUM method for protein sequence score matrices, upon which the glycan version is based. In place of amino acids, the concept of *links* was defined as two monosaccharides and their glycosidic linkage, which includes the full linkage information (carbon numbers and conformation) as well as the monosaccharide names.

Another issue lay in the selection of glycans to use in computing the score matrix, since the concept of families of glycans based on conserved motifs did not exist. In order to determine the glycans to be used, glycan families could be defined computationally and/or generated based on the classic classification of glycans, which is derived from the core structure and/or determined by the conjugate to which the glycan is bound (e.g., lipids and GPI-anchors) (as introduced previously in Section 1.3). In this work, a computational approach based on the classic glycan families was taken to select the appropriate glycans. First, the overall distribution of glycans in the KEGG GLYCAN database according to class was found. Here, glycan size, defined as the number of sugars in the structure became an issue. On average, depending on the glycan class, glycans consisted of approximately seven sugars, ranging from three to up to 15 on average. Comparing a glycan of size three to one of size 15 would not produce a meaningful alignment. In order to avoid this, glycans of size less than five were first removed from the dataset. Additionally, sufficiently large glycan classes of at least 500 representative glycans were selected, resulting in a selection of the three classes of N-linked, O-linked, and sphingolipid glycans. Finally, blocks needed to be defined for each class. A hierarchical clustering of the all-by-all local exact match scores of the selected glycans was performed, and groups of glycans consisting of approximately 200 structures each were extracted as the final glycan data set.

In the matrix calculation, the pairwise alignment of each pair of glycans in each block was used to compute the frequency of alignment of pairs of links, denoted as f_{ij} for links i and j. The probability of occurrence of aligning i and j, denoted by q_{ij} was then computed by dividing f_{ij} by the total number of alignments. Next, the probability that link i was aligned was calculated by

$$p_i = q_{ii} + \sum_{i \neq j} \frac{q_{ij}}{2}.$$

Finally, the expected probability of aligning i and j was computed as:

$$e_{ij} = \begin{cases} p_i p_j & \text{for} \quad i = j \\ 2 p_i p_j & \text{for} \quad i \neq j \end{cases}$$

As a result, the score for aligning links i and j could be computed by

$$s_{ij} = \log_2 \frac{q_{ij}}{e_{ij}}.$$

Thus, the glycan score matrix contained the log odds score of the expected frequency of alignment of link pairs (Aoki et al. (2005)). From this matrix,

TABLE 4.1:   Difference in ranking
using glycan score matrix. Asterisks
indicate that the glycan belongs to the same
glycan class as the query.

Ranking	Without matrix	With matrix
1	G00086	G04134*
2	G00192	G04072*
3	G04134*	G05073*
4	G04906*	G04906*
5	G00407*	G05305*
6	G00975	G04140*

those links that are positioned similarly, and thus those that are potentially
"functionally" similar could be analyzed. This matrix could also be used to
improve the KCaM algorithm to produce more biologically meaningful results.
As an example, an *O*-linked glycan (Figure 4.1) was used as a query to KCaM
with and without the use of the glycan score matrix.

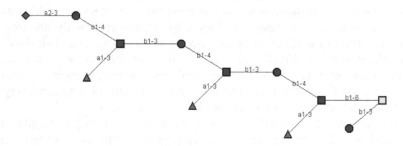

FIGURE 4.1:   Glycan structure used to test the glycan score matrix.

The resulting ranking of glycans with and without the score matrix is listed
in Table 4.1. In this table, glycans that belong to the same class as the query
glycan are indicated by asterisks next to their glycan IDs. Obviously, the top
scoring glycans using the score matrix are more biologically relevant glycans
due to the fact that they belong to the same glycan class. The structures for
the top ranking structures G00086 and G04134 are displayed in Figure 4.2.

4.2.3   Glycan variation map

Until glycan structure databases were developed and organized, no one re-
ally knew the breadth in variety of the glycan structures that were known.
Thus, with the creation of the KEGG GLYCAN database (Section 3.1.1),

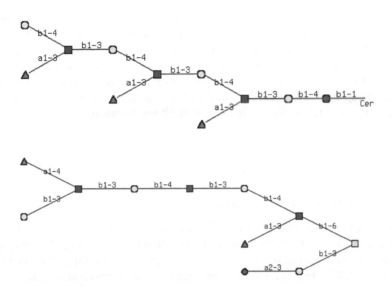

FIGURE 4.2: Resulting structures from testing the glycan score matrix on the query structure in Figure 4.1. The top structure is the top ranking structure without using the matrix, and the bottom structure is the top result using the matrix.

a glycan variation map was generated, consisting of the merged structures of all known glycans in the glycome (registered in the database). This map was built by utilizing the KCaM algorithm to align every pair of glycans and merging the aligned links together. That is, given two glycans A and B, the links of B that align with A are preserved while those that do not align are merged into A, resulting in a tree that is the union of A and B, which can be called AB. By repeating this next with AB and each glycan in the glycome, an extremely large structure representing the entirety of the glycome can be obtained. Such a structure was built based on the glycans that contained the same residue at the reducing end. In these structures, it was found that the variety of glycans compared to what would be expected in theory was quite limited (Hashimoto et al. (2005)). The most varied variation map was that of glycolipids, containing glycans with glucose residues at the reducing end. The number of branches for the variation map of these structures totalled a little over 1000. For the N-glycan class of structures having a GlcNAc at the reducing end, a total of over 750 branches were obtained in the variation tree. Considering that for a single link there are at least eight possible glycosidic linkage variations for any two residues and that there are over 10 different residues in mammals, there are theoretically at least 800 possible combinations of residues and linkage conformations for a single link alone. Since a single N-glycan structure may contain on average 10 links, approximately 8000 branches are theoretically possible in the variation tree. However, only one-tenth of this number for all known *N*-linked glycans in KEGG were found. Thus, it may be assumed that nature does not require all possible glycan structures for mammalian systems to function (or they have not yet been registered). In either case, this number was surprisingly small for all the knowledge that has been accumulated for mammalian systems. Bacterial systems have yet to be comprehensively analyzed as such.

4.3   Bioinformatic methods

The bioinformatic methods are differentiated in this section from algorithmic methods in that biological problems are directly addressed. In particular, genomic information related to the enzymes that interact with glycans, three-dimensional (3D) structures of glycans, and systems-level analysis of glycans are studied.

4.3.1   Glycan structure prediction from glycogene microarrays

The characterization of glycan structures from biological samples is one of the biggest challenges in glycomics today. Mass spectrometry (MS) and nuclear magnetic resonance (NMR) techniques have enabled faster and more accurate characterization of glycans compared to just 10 or 20 years ago. However, it has not yet reached the stage that shotgun sequencing has reached. Thus, as an alternative, methods to perform glycan structure characterization by predicting glycan structures in a particular cell through glycosyltransferase gene expression profiles was developed (Kawano et al. (2005)). In this method, the concept of a co-occurrence score was used based on the frequency of occurrence of pairs of links (defined as two linked monosaccharides and their glycosidic linkage information) within the same glycan structure. That is, for any glycan structure, all the links within that structure regardless of their position would be considered to co-occur with one another. Then for each pair of links i and j, the following correlation coefficient was computed.

$$S_p(i,j) = \frac{\sum_k (x_i(k) - \bar{x}_i)(x_j(k) - \bar{x}_j)}{\sqrt{\sum_k (x_i(k) - \bar{x}_i)^2 \sum_k (x_j(k) - \bar{x}_j)^2}}$$

where $x_i(k)$ indicates the number of times link i appears in glycan k, and $\bar{x}_i$ indicates the average number of appearances of link i across all glycans.

These correlation coefficients were summarized in a so-called "score matrix" for every pair of links, developed with the expectation that the substrate specificity of glycosyltransferases could be compactly captured. Once this co-occurrence score matrix was developed, it could be used to make predictions from expression data of glycosyltransferases which coexpress. The expression data was transformed into binary data where only those genes whose expression values greater than a specific threshold would be taken into consideration. The genes with sufficient expression would then be translated into the link that they catalyze, and the database of glycans would be scored according to those expressed genes and their co-occurrence score. That is, the existing database of glycan structures could be scored based on the glycosyltransferase expression data from microarray experiments, for example, to determine the most likely glycan structures to be synthesized. Thus, given a list of p expressed glycosyltransferases with expression values $\{z_1, z_2, \ldots, z_p\}$ converted to a list of links $\{q_1, q_2, \ldots, q_m\}$, the following scoring function could be computed for a particular candidate glycan g:

$$S(g) = \frac{1}{\sum_{i=1}^{m} I(q_i \in g)} \sum_{j=1}^{p} \sum_{k=1}^{m} I(z_j = 1) I(q_k \in g) S_p(q_j, q_k)$$

where $q \in g$ refers to whether link q exists in glycan g, $I(x)$ is an indicator function returning 1 if x is true and 0 if x is false. As a result, the existing glycans in the database that obtained the highest scores would be predicted as the most likely candidates to be synthesized by the given genes.

Since this method depended on the co-occurrence matrix which was generated based on existing data, it was quite biased towards the database at hand. Therefore this method was further improved such that (1) the database of glycans was augmented with new glycans that should theoretically exist, and (2) the prediction score computation was modified such that the expression values were used directly as opposed to using binary values based on a threshold. The first step was performed by scanning the glycan database and finding those that differed by more than one link. That is, considering that glycosyltransferases typically catalyze only one link at a time, if two similar glycans in the database existed but differed by more than one link, then "intermediate" glycans that in the process of synthesizing the larger structure should also exist. Thus these "intermediate" glycans were added to the database to generate the score matrix as well as to predict glycan structures. Figure 4.3 is a schematic of how a new entry can be presumed to exist and thus added to the database based on the existing entries Entry 1 and 2. With this augmented database, better scoring results can be expected. The second step is the modification of the scoring system. Instead of binary values correlated with expression determined by some threshold, the expression values themselves are used in computing the score, as follows.

$$S(g) = \frac{1}{\sum_{i=1}^{m} I(q_i \in g)} \sum_{j=1}^{p} \sum_{k=1}^{m} z_j I(q_k \in g) S_p(q_j, q_k)$$

using the same variables as in the earlier equation.

For verification, a dataset of glycans related to acute lymphocytic and mye-
locytic leukemia were used, and those structures containing Lewis^a, Lewis^x or
sialyl-Lewis^x epitopes, which are known to be related to cancer, were ranked.
As a result, these cancer markers were found to be ranked more highly compared to the original method. Furthermore, the newly-added glycan entries
were also found to be ranked highly in the results (Suga et al. (2007)). This
method has been applied and implemented as a tool called GECS (Gene Expression to Chemical Structure), which is available at http://www.genome.jp/
tools/gecs/.

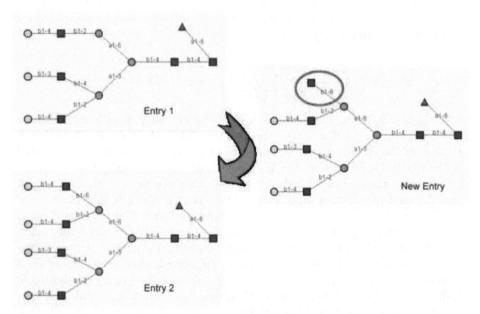

FIGURE 4.3: Schematic of generating intermediate glycans as used by
GECS.

4.3.2   Glyco-gene sequence and structure analysis

The term glyco-gene refers to those genes that interact with glycans, including glycosyltransferases, lectins, and sugar-nucleotide synthetases and transporters. The sequence analysis of lectins has resulted in the identification of
carbohydrate-recognition domains (CRDs) which are specific for each lectin
subclass. These have been introduced in Chapter 2.2.

An early work on classifying glycosyltransferases based on amino acid sequence was able to find many conserved residues among biochemically similar

enzymes, but in some cases there was no sequence homology in glycosyltransferases that were assumed to be in the same family (Campbell et al. (1997)). Thus it was concluded that it was necessary to perform structural analyses to better classify glycosyltransferases. The CAZy database, introduced in Section 3.2.4, was developed as a result of such structural analyses on not only glycosyltransferases but also on other glycan-related enzymes such as glycosidases and hydrolases. All of the enzymes in CAZy could be classified as having one of two types of conformations called GT-A and GT-B.

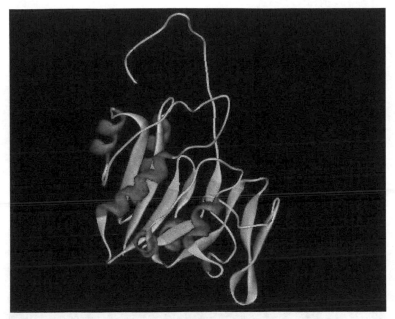

FIGURE 4.4: 3D structure of human glucuronyltransferase, GlcAT-P (PDB ID: 1v82), which takes on a GT-A fold.

An example of the GT-A fold is given in Figure 4.4, which is the 3D structure of human glucuronyltransferase, GlcAT-P. The GT-A fold can be described as two $\beta/\alpha/\beta$ domains that tend to form a continuous central sheet of at least eight β-strands, which some authors describe as a single domain fold. This structure contains an N-terminal nucleotide binding domain and a C-terminal acceptor-binding domain. The N-terminal domain often contains the DxD motif, which is known to interact with nucleotide donors. In contrast, the GT-B fold also displays two Rossman-like $\beta/\alpha/\beta$ domains, but these associate less tightly and are separated by a deep cleft. An example of this fold is given in Figure 4.5, using ADP-heptose LPS heptosyltransferase II from *E. coli*. In more recent work, a third family named GT-C was found

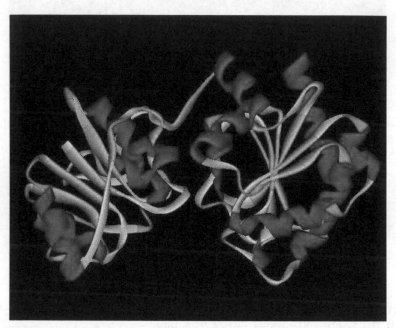

FIGURE 4.5:   3D structure of E. coli ADP-heptose lps heptosyltransferase II (PDB ID: 1psw), which takes on a GT-B fold.

TABLE 4.2: Classification of glycosyltransferases by profile HMM

Classification	CAZy families	Pfam IDs	PDB IDs
GTS-A	2, 6, 7, 8, 12, 13, 14, 16, 21, 24, 25, 27, 31, 32, 40, 43, 44, 45, 46, 47, 48, 49, 54, 55, 60	PF00535, PF003414, PF02709, PF01501, PF03071, PF02485, PF01755, PF01762, PF03360, PF03016, PF02364	1h71, 1fg5, 1fgx, 1g9r, 1l10, 1f08, 1fgg
GTS-B	1, 3, 4, 5, 9, 19, 20, 28, 30, 33, 35, 41	PF00201, PF00862, PF01075, PF02684, PF00982, PF04101, PF00343	1llr, 1f0k, 1ahp, 1em6, 1a8i, 1ygp
GTS-C	22, 39, 50, 53, 57, 58, 59	PF02366, PF03155	
GTS-D	11, 23, 37	PF01531	
Unclassified	10, 15, 17, 18, 26, 29, 34, 35, 38, 42, 51, 52, 56	PF00852, PF01793, PF0383, PF00777, PF00912	

in a bacterial sialyltransferase (Breton et al. (2006)).

In all cases, it is known that a DxD motif is important for ligand binding, but since it is so short, it is often difficult to detect at a significant level. Thus the profile hidden Markov model (HMM) was used in an attempt to capture patterns from the sequences of glycosyltransferases (Kikuchi and Narimatsu (2003)). An introduction to HMMs and profile HMMs is provided in Appendices B.2 and B.2.4. It was assumed that this model would be more robust to noisy or less conserved data. In this work, each member of a single CAZy family was multiply aligned using ClustalW and then edited manually. These sequences were then used as the input to the profile HMM program. For large families, 10 members were selected for the alignment, and after the profile was generated, the rest of the members were aligned. Finally, to classify all the sequences based on the generated profiles, each glycosyltransferase sequence was used as a query against the profiles and were clustered into the same superfamily if the profile score produced a significance value below 0.001. As a result, it was proposed that the glycosyltransferases could be categorized into five groups, named GTS-A, GTS-B, GTS-C, GTS-D and "unclassified." The resulting classification is provided in Table 4.2.

4.3.3 Glyco-related pathway analysis

The main biosynthetic pathways for *N*-linked glycans from the common dolichol-linked glycan have been well established in mammalian cells, as in-

TABLE 4.3: Enzymes considered in GlycoVis.

No.	Name	Symbol	Substrate specificity
1.	Mannosyl-oligosaccharide 1,2-α-mannosidase	Man I	Free α(1,2)Man
2.	Mannosyl-oligosaccharide 1,3-1,6-α-mannosidase	Man II	Free α(1,3) or α(1,6)Man with opposing β(1,2)GlcNAc
3.	α-1,3-mannosyl-glycoprotein 2-β-N-acetylglucosaminyltransferase	GnT I	Man5
4.	α-1,6-mannosyl-glycoprotein 2-β-N-acetylglucosaminyltransferase	GnT II	β(1,2)GlcNAc must be added to α(1,3)Man branch first; no bisecting GlcNAc; inhibited if β(1,4)Gal is already added to α(1,3)Man
5.	β-1,4-mannosyl-glycoprotein 4-β-N-acetylglucosaminyltransferase	GnT III	β(1,2)GlcNAc must be added to α(1,3)Man branch first; inhibited by any added Gals
6.	α-1,3-mannosyl-glycoprotein 4-β-N-acetylglucosaminyltransferase	GnT IV	β(1,2)GlcNAc must be added to α(1,3)Man branch first; core α(1,6)Fuc required; inhibited if β(1,4)Gal is already added to α(1,3)Man; no bisecting GlcNAc
7.	α-1,6-mannosyl-glycoprotein 6-β-N-acetylglucosaminyltransferase	GnT V	β(1,2)GlcNAc must be added to α(1,6)Man branch; core α(1,6)Fuc required; inhibited if β(1,4)Gal is already added to α(1,6)Man; no bisecting GlcNAc
8.	Glycoprotein 6-α-L-fucosyltransferase	FucT	At least one GlcNAc added; no bisecting GlcNAc; inhibited if all terminal residues are capped with β(1,4)Gal
9.	β-N-acetylglucosaminylglycopeptide β-1,4-galactosyltransferase	GalT	Free GlcNAc on any branch
10.	β-Galactoside α-2,3/6-sialyltransferase	SiaT	Free β(1,4)Gal on any branch

troduced in Section 1.4 and Kornfeld and Kornfeld (1985). Thus the major enzymes involved in this process are known. Based upon this information, several groups have developed tools for the analysis of the *N*-linked glycan biosynthesis pathway. These are introduced in this section.

4.3.3.1 GlycoVis

Because the biosynthetic pathways of *N*-linked glycans on glycoproteins involve a relatively small number of enzymes and nucleotide sugars mainly localized in the Golgi apparatus, a computational program to predict and visualize the relevant pathways up to a particular *N*-linked glycan structure was developed, called GlycoVis (Hossler et al. (2006)). This tool also displays the glycan structure distribution of the reaction paths using different colors. The enzymes and their substrate specificity rules that were used in this model are listed in Table 4.3.

N-glycan biosynthesis is not a straightforward process. For example, it is known that the same glycan structures may serve as substrates for multiple enzymes, and different enzymes may produce the same glycan structures. Different glycosidic linkages may also be produced by the same enzyme depending on the substrate. Thus, a graph, or network, of the biosynthesis pathway, where nodes represent glycans and edges representing enzymatic reactions, may become quite complex. Using the enzymes in Table 4.3 and starting from the Man9 structure, 341 distinct *N*-glycans were produced.

A relationship matrix was used to model this network, where each row corresponds to a glycan reactant, and each column to a glycan product. The cells of the matrix are numbered with 0 if no enzyme can catalyze the product from the reactant, and with numbers 1 to 10 for each of the enzymes in Table 4.3. This relationship matrix is used to visualize the entire network. Given a table of distributions of glycans (e.g., 10% glycan X, 30% glycan Y and 60% glycan Z), the program computes the possible reaction pathways producing the given glycans starting from the Man9 structure. Colors are used in the visualization to indicate the percentage of the total glycan input. Since not all of the glycans may be completely characterized, similar glycan structures are grouped together. Thus GlycoVis distinguishes between uniquely determined glycan structures and those not uniquely determined by representing the former by an ellipse shape and the latter by a box shaped node. The edges are highlighted according to the color of their parent nodes. Thus the display of glycans from experimental observation or model prediction can be made onto pathway maps to assist in understanding the possible reaction paths used to lead to the given glycans.

Several test data sets were used to evaluate GlycoVis. The glycan distribution in Chinese hamster ovary (CHO) cell derived tissue plasminogen activator (TPA) was evaluated according to the literature, and as a result, explanations (i.e., the lack of Man 1 and GnT I) could be hypothesized for the accumulation of high mannose glycans in these cells. In addition, the *N*-glycan microheterogeneity in human vs. mouse IgG was visualized and compared using the distribution as reported in Raju et al. (2000). Ignoring sialylated structures due to their small amounts, eight enzymes were considered in the biosynthesis process since neither GnT IV nor GnT V were shown to be active towards glycans synthesized on IgG molecules (Mizuochi et al. (1982)). The disparity between these pathways revealed differences in their *N*-glycan processing. For example, many ungalactosylated glycans were found, suggesting that more GalT could be added in order to increase the level of terminal glycan processing.

As an application of GlycoVis, a systems analysis of *N*-glycan processing in mammalian cells was performed (Hossler et al. (2007)). A mathematical model of glycan biosynthesis in the Golgi was developed, and the various reaction variables on the resulting glycan distribution were analyzed. The Golgi model was modeled as four compartments in series, and the mechanism of protein transport across the Golgi was modeled in two ways: vesicular transport

TABLE 4.4:   Coding scheme used by Krambeck and Betenbaugh (2005) for representing *N*-linked glycans in a compact form.

Number	Name	Description
1	Man	Number of mannose residues
2	Fuc	Number of core fucose residues (up to one)
3	Gnb	Number of bisecting GlcNAc residues (up to one)
4	Br1	Extension "level" of branch 1 (see Table 4.5)
5	Br2	Extension "level" of branch 2 (see Table 4.5)
6	Br3	Extension "level" of branch 3 (see Table 4.5)
7	Br4	Extension "level" of branch 4 (see Table 4.5)
8	Gal	Number of galactose residues
9	Sia	Number of sialic acid (NeuAc) residues

Man, Fuc, Gal and Sia refer to the number of the given monosaccharide residue. Gnb is the bisecting GlcNAc, and Br1, Br2, Br3, Br4 refer to each of the four branches.

and Golgi maturation, which resemble four continuous mixing tanks (4CSTR) and four plug-flow reactors (4PFR) in series, respectively. Comparing the glycans profiles predicted by these two models, it was found that the 4PFR system was more likely, despite the fact that it has been surmised that the true model is actually a composite between the PFR and CSTR models (Elsner et al. (2003); Mironov et al. (1998)). To assess this model, it was demonstrated that, with a sufficient holding time in the Golgi compartments and by the spatial localization of enzymes to specific compartments, all terminally processed *N*-glycans could be synthesized as homogeneous products.

4.3.3.2   A mathematical model for *N*-glycan biosynthesis

One of the first purely mathematical models of *N*-glycan biosynthesis was developed by Krambeck and Betenbaugh (2005), as an extension of Umana and Bailey (1997). In this work, glycan structures are represented as a list of nine numbers (hereafter referred to as the GlycanCode) indicating the number of monosaccharides and branching configuration. Table 4.4 describes the meaning of each of the nine numbers in the GlycanCode. Man, Fuc, Gal and Sia refer to the number of the corresponding monosaccharide residue to add. Gnb is the bisecting GlcNAc, and Br1, Br2, Br3, Br4 refer to the extension level of each of the four branches. Table 4.5 describes the extension levels and the structures corresponding to each level.

Using this glycan encoding scheme, the list of enzymes shown in Table 4.6 was used to represent substrate specificity and the linkage to catalyze. This table lists the set of mathematical rules that the acceptor structure must satisfy in order for the given enzyme to add the corresponding linkage.

To describe the biosynthesis process, let us take the enzyme GalT on the structure in Figure 4.6a as an example. The GlycanCode for this structure is 400001000. Since branch 3 in this structure is 1, the third substrate rule for

TABLE 4.5: Extension levels as used by Krambeck and Betenbaugh (2005) for representing the terminal extensions of branches.

Level	Structure
1	■
2	○—b1-4—■
3	◆—a2-3—○—b1-4—■
4	■—b1-3—○—b1-4—■
5	○—b1-4—■—b1-3—○—b1-4—■
6	◆—a2-3—○—b1-4—■—b1-3—○—b1-4—■

TABLE 4.6: Table of enzymes and substrate rules used by Krambeck and Betenbaugh (2005).

Enzyme	Rule	Activity
Man I	Man>5	Man:-1
Man II	Man>3 & Br4=1 & Gnb=0	Man:-1
FucT	Fuc=0 & Br4>0 & Gnb=0 & Gal=0	Fuc:+1
GnT I	Br4=0 & Man=5	Br4:+1
GnT II	Br2=0 & Man=3 & Br4=1 & Gnb=0	Br2:+1
GnT III	Gnb=0 & Br4>0 & Gal=0	Gnb:+1
GnT IV	Br3=0 & Br4=1 & Gnb=0	Br3:+1
GnT V	Br1=0 & Br2=1 & Gnb=0	Br1:+1
GnT E	Br1=2	Br1:+2
GnT E	Br2=2	Br2:+2
GnT E	Br3=2	Br3:+2
GnT E	Br4=2	Br4:+2
GalT	Br1=1 or Br1=4	Br1:+1 & Gal:+1
GalT	Br2=1 or Br2=4	Br2:+1 & Gal:+1
GalT	Br3=1 or Br3=4	Br3:+1 & Gal:+1
GalT	Br4=1 or Br4=4	Br4:+1 & Gal:+1
SiaT	Br1=2 or Br1=5	Br1:+1 & Sia:+1
SiaT	Br2=2 or Br2=5	Br2:+1 & Sia:+1
SiaT	Br3=2 or Br3=5	Br3:+1 & Sia:+1
SiaT	Br4=2 or Br4=5	Br4:+1 & Sia:+1

Man, Fuc, Gal and Sia refer to the number of the given monosaccharide residue to add. Gnb is the bisecting GlcNAc, and Br1, Br2, Br3, Br4 refer to the extension level of each of the four branches. Thus the Activity number corresponds to the number to add to the corresponding GlycanCode.

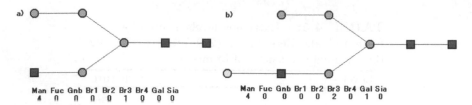

Man	Fuc	Gnb	Br1	Br2	Br3	Br4	Gal	Sia
4	0	0	0	0	1	0	0	0

Man	Fuc	Gnb	Br1	Br2	Br3	Br4	Gal	Sia
4	0	0	0	0	2	0	1	0

FIGURE 4.6: a) Example structure of an *N*-glycan and its GlycanCode. b) Resulting structure from enzyme GalT.

GalT is satisfied in Table 4.6. Thus the values of Br3 and Gal are increased by one each, resulting in GlycanCode 400002010, which corresponds to the structure in Figure 4.6b. Linkage information is omitted for simplicity.

Using this enzyme table, any basic *N*-glycan structure could be specified in GlycanCode format and catalyzed structures could be predicted. In this work, the high-mannose Man9 structure was used as a starting structure, and applying all enzymes in Table 4.6, 7,565 structures and 22,871 reactions were generated. This model was further augmented with rate parameter adjustment rules to account for enzyme kinetics. As a result, it was shown that the distribution of structures predicted by the model could be correlated with known results from the literature.

4.3.4   Mass spectral data annotation

The automatic annotation of mass spectrometry data of glycans and glycoproteins is one of the major bottlenecks in glycomics. The manual process of such annotations may take weeks, if not months. Thus several methods and tools have been developed to ease this process. Those that have been developed and made freely available are introduced in Section 4.5.6 while the methods used by those that are not available are introduced here. A brief introduction to the relevant mass spectrometry techniques is provided in Appendix C.

4.3.4.1   Cartoonist

The Cartoonist algorithm performs automatic annotation of *N*-glycans in MALDI-TOF spectra (Goldberg et al. (2005)). In general, this algorithm consists of three main elements: (1) Restriction of annotations to a library of approximately 2800 biosynthetically plausible structures, which were constructed from about 300 manually-specified archetype structures. (2) Determination of the precision and calibration of the instrument used to generate the spectrum, performed automatically based on the spectrum itself. (3) Assignment of a confidence score to each identified structure. Note that in this tool, linkage information is not considered.

In part (1), the structure library starts with 300 reference N-glycans all containing a core Fuc, no bisecting GlcNAc, and sialic acid residues represented as NeuAc. These structures were derived from knowledge of the N-glycan biosynthetic pathway (Lowe and Marth (2003)). Then three rules are used to generate other N-glycans with variations. The first rule generates glycans without a core Fuc. The second rule generates bisecting GlcNAcs automatically, and the third rule systematically replaces NeuAc with NeuGc. With these three rules, 2500 additional N-glycan structures were generated. It is noted that the glycan library can contain multiple structures for the same atomic number. These are discriminated by ranking them according to knowledge about the sample, such as by giving LacdiNAc (GalNAcβ1-4GlcNAcβ) a lower rank by default, but raising it for samples known to be rich in LacdiNAc structures. Furthermore, demerits are given to structures containing specific residues, including terminal GlcNAcs that are not modified by a Gal, LacdiNAc moieties, sialyl-Lewis^x moieties, disialic acid structures, antennae containing multiple fucose residues, hybrid structures, and structures with five antennae. It is also noted that multiple structures may correspond to a similar mass or alternative topologies of a glycan. Using this library of structures, theoretical spectra can be generated for each structure, which is used to match against the observed spectrum.

In part (2), the calibration of a spectrum is determined by first finding approximately 15 high-confidence peak assignments, which are peaks that are relatively large, have isotope envelopes that closely match their theoretical values, and have a mass very close to that of the predicted glycan. For each peak of mass m_i, the deviation between theoretical and observed peak masses is computed as $d_i = pred - obs$. The pairs (m_i, d_i) over all i are fitted to a line $\alpha m + \beta$ to obtain corrected deviations $d_i' = d_i - (\alpha m_i + \beta)$. Taking into account outliers generated from incorrect assignments, the best linear fit is found (Fischler and Bolles (1981)). This process is repeated by finding a new set of high confidence peak assignments to refine the estimate of d'. As a result, the values of α and β are used as a calibration of the instrument, and the spread of the deviations d' is the precision. Thus, given a spectrum to annotate, these values are first computed such that the matching of the theoretical and observed peaks can be normalized.

Finally, in part (3), a confidence value is computed for each peak to represent the likelihood of a particular assignment. This value is calculated by comparing the deviation to the standard deviation σ and the observed isotope envelope to the predicted envelope. That is, for a peak of mass m, the closeness of the deviation is computed as d'/σ. In addition, the height h_i of the peaks at mass $m - 1$, m, $m + 1$, $m + 2$, and $m + 3$ are compared to the sequence $0, f_0, f_1, f_2, f_3$, where f_k is the predicted abundance of the $+k$ isotope of the associated glycan. These values for the height and frequencies can be represented as vectors, which are then normalized. Subsequently, the norm of their difference can be computed as a measure of the match between theoretical and observed isotope envelopes. The confidence value is the sum of the deviation and vector difference norm, scaled such that a peak with a deviation of 1σ and a vector difference of 1σ has a score of 10, with lower scores indicating assignments of higher confidence.

4.3.4.2   CartoonistTwo

The Cartoonist algorithm was next extended to handle the annotation of O-glycan structures from fragmentation spectra. This algorithm was applied to data obtained from FT-ICR (Fourier Transform-Ion Cyclotron Resonance) mass spectrometry, employing multiple rounds of SORI-CID (Sustained Off-Resonance Irradiation Collision-Induced Dissociation) or IRMPD (Infrared Multiphoton Dissociation) fragmentation. This tool, called CartoonistTwo (Goldberg et al. (2006)), extends the scoring algorithm of the previous method by incorporating the following points:

1. The use of low-intensity peaks excluding noise peaks by utilizing a statistical confidence score based on both intensity and m/z.

2. The assumption of a low-energy glycan fragmentation, where single monosaccharides are removed from the glycan one at a time. In this case, the charge remains with the larger daughter ion.

3. The use of peaks that are not present in the spectrum, in addition to the present ones. These can be considered fragments of a proposed structure that does not appear in the spectrum.

The CartoonistTwo program consists of three steps: (a) the processing of spectra by picking peaks that are likely to represent glycans and the assignment of probability values to them, (b) an enhanced version of Cartoonist, which tentatively assigns possible glycan composition to the peaks and recalibrates m/z measurements based on these assignments, and (c) the identification and scoring of candidate glycan structures.

In step (a), a peak histogram is generated from the thousands of peaks output from the MS instrument. Then, approximately 15 (x, y) points are plotted, where x is the center intensity of a histogram bin and y is the logarithm of the number of peaks within that bin. This plot is fitted to a quadratic equation $-a_0 - a_1 x - a_2 x^2$, which can then be used to compute the probability that a noise peak has intensity at least α by:

$$p(\alpha) = \int_{\alpha}^{\infty} e^{-(a_0 + a_1 x + a_2 x^2)} \, \mathrm{d}x = \frac{1}{\sqrt{a_2}} e^{-a_0 + a_1^2/4a_2} \mathrm{erfc}((\alpha + \frac{a_1}{2a_2})\sqrt{a_2})$$

Given N peaks in the spectrum, then at least one noise peak of intensity $> \alpha$ can be expected to occur with probability $1 - (1 - p(\alpha))^N$. Thus the significance of a peak of intensity α can be set to $(1 - p(\alpha))^N$. For the recalibration in step (b), peaks with sufficiently high p values are matched to glycan masses including those with single water losses. A robust statistical regression method

TABLE 4.7: Scoring results of CartoonistTwo. Refer to the text for explanations of the scoring functions and performance evaluation.

score function	# correct	# ties	# second	# misses	performance
F1	7	27	0	0	0.502
F2	9	24	1	0	0.514
F3	19	3	8	4	0.716
F4	20	3	8	3	0.730
F5	20	3	8	3	0.732

is used to compute a correction curve mapping measured masses to theoretical masses, which can then be used for recalibration. CartoonistTwo next models the recalibrated mass errors arising from a normal distribution and gives each peak a confidence value, which is computed by multiplying the probability density at the peak's mass error by its significance. Thus, this confidence value gives the probability that the peak is indeed a glycan fragment.

For a sequence of MS^n spectra, CartoonistTwo first sets peak significances and recalibrates mass measurements for each spectrum individually. It then takes the union of all the significant peaks found in all spectra. If a peak is observed more than once, it is assigned the maximum of its confidence values.

Finally, in step (c), candidate glycan structures are given scores based on peak confidences. This scoring function could be improved by incorporating point 2 above, where a small bonus is applied for each observed fragment with a path of observed peaks to the original glycan structure. Alternatively, incorporating point 3, fragments of the candidate that are not observed could be penalized. Combinations of these improvements are also possible.

In the original work, these scoring functions were evaluated against 34 sequences of SORI-CID MS^n spectra which were annotated manually. Table 4.7 lists the results of using the original scoring function (F1), F1 plus point 2 (F2), F1 plus point 3 (F3), F1 plus both points 2 and 3 (F4), and F4 but with multiple penalities in proportion to the number of fragments of that mass that are unobserved (F5). In this table, the performance is evaluated by computing $\mu = 1/r$, where r is the rank that the scoring function assigned to the correct topology when there are no ties. In the case of ties, it is assumed that the tied structures appear in random order, and so r becomes the expected rank of the correct structure over all possible random orders. From these results, it is apparent that the scoring function including both improvements was able to annotate the spectra most accurately. In fact, it matched the known answer in over half of the spectra and even caught errors in the manual annotation.

4.3.4.3 Peptoonist

Peptoonist is an annotation program for identifying N-glycopeptides from a series of both liquid chromatography (LC)-MS and MS/MS spectra, obtained from glycoprotein samples (Goldberg et al. (2007)). This program consists of four steps: (1) recalibration, (2) quality filtering of single-MS peaks, (3), MS/MS scoring, and (4) glycopeptide annotation of single-MS peaks.

In step (1), Peptoonist uses peaks from unmodified peptides known to fragment well for the recalibration of both MS and MS/MS spectra. For all identified MS/MS spectra, the peptide and retention times are used to find matching single-MS peaks of any charge. Each of these matches are plotted as (x, y), where x is the m/z of the single-MS peak, and y is the difference between the theoretical and actual m/z of the peptide. Using these points, recalibration is performed by using a robust line-fitting algorithm (Bern and Goldberg (2005)), whose resulting recalibration lines are used to recalibrate all the single-MS spectra in the data set. Recalibration of the MS/MS spectra is performed based on the observed B and Y ions, whose resulting recalibration lines are averaged to recalibrate all the MS/MS spectra.

In step (2), the single-MS spectra are searched for series of peaks that match the isotope ratios of glycopeptides by fitting a theoretical set of Gaussians (using approximately 6000 glycan compositions generated by Cartoonist and a set of peptides, consisting of either all tryptic peptides or those found in the MS/MS analysis) to the actual peak enveleope. This fitting determines the monoisotopic mass and the charge. Those peaks that fit within a certain threshold are selected. This threshold is determined based on the scatter of (x, y) points. Note that at the time of this writing, Peptoonist assumes that peak envelopes do not overlap. Regardless, this fitting step is the most time-consuming due to the computation of the quality of the fit in addition to the fitting computed above. This quality computation uses the distance $d(G, (x, y))$ between (x, y) and the Gaussian curve G, estimated by dividing $(y - G(x))$ by the square root of $1 + G'(x)$. The total distance over the points in the envelope are then summed to compute $D = \sum d(G, (x_i, y_i))$. Since the fit quality may be insignificant between the best (D_1) and second-best (D_2) fits, the total fit quality score is computed as $Q = D_1 + 4(D_1/D_2)$.

Using the selected peaks from step (2), in step (3), each MS/MS spectrum corresponding to a selected single-MS envelope is scored against all possible theoretical glycopeptides, and the best score is returned. This score is computed by searching the spectrum for three types of theoretically generated fragment peaks: (i) B or Y ions from the base peptide, (ii) glycan fragments, and (iii) the entire peptide plus a glycan fragment. Since a peak of type (ii) must be one of the top 10 due to the nature of the data, the ranks of the peaks of types (i) and (iii) are scored using the following scoring function and summed: $1/(1 + (rank/100)^2)$.

Finally, in step (4), a score is assigned to each envelope using the logistic function $1/1(+0.2e^Q)$, which drops to half its maximum at $Q = 4$. Similarly, if the actual error is ϵ and the maximum expected error is E, then the logistic function becomes $1/(1 + 0.25e^{1.8(\epsilon/E)})$, which drops to half its maximum at $\epsilon/E = 1$. These functions are combined in the following manner in order to score each envelope: $1/(1 + 0.2e^Q + 0.25e^{1.8(\epsilon/E)})$. Then the final score for each glycopeptide identified is computed as the weighted sum of scores from envelopes assigned to that glycopeptide and all related glycopeptides.

To validate the Peptoonist algorithm, a mixture of mouse zona pellucida proteins ZP2 and ZP3 were used as the biological sample. The spectra obtained from this sample was annotated manually, resulting in 58 different glycoforms that were attached to Asn[273] of the ZP3 sequence [257] PR-PETLQFTVDVFHFANSSR [276] having mass 2347.18. This result was confirmed by Peptoonist, which also found 55 of the 58 glycoforms annotated manually, in addition to another 54 glycoforms. In fact, two of the three missed glycoforms were found to be rather weakly supported.

4.3.4.4 GLYCH

In contrast to the previous methods, a program called GLYCH (GLYcan CHaracterization) was developed to annotate MS/MS spectra of oligosaccharides using a dynamic programming technique (Tang et al. (2005)). GLYCH represents the n monosaccharide residues of a glycan by $r_1, r_2, \ldots, r_n$, where r_n is the root node and $r_i \in \mathcal{R}$. Glycosidic linkages are represented as $b_i \in \mathcal{B}$ for the linkage between residue r_i and its parent. It also defines a prefix residue mass (PRM) m_i as the total mass of residues in the subtree rooted by residue r_i. Thus m_n is the total mass of the glycan. In order to account for glycosidic linkages, a trimer $F_i = (m_i, r_i, b_i)$ was defined, called the prefix residue feature (PRF), which consists of the PRM m_i, the monosaccharide r_i and the linkage type b_i of the residue to its parent.

Given an MS/MS spectrum, it is assumed that every peak should correspond to one of the fragmentation ions B/Y, C/Z or A/X, unless it is a noisy peak. Assuming also that one peak p is the fragmentation ion type j of residue r with linkage b, the corresponding PRM m_{jr} could be computed as:

$$m_{jr} = m_p + \delta m_{jr} \quad \text{for} \quad j = B, C, {}^{0,2}A, \ldots, {}^{3,5}A$$
$$m_{jr} = m_n - (m_p + \delta m_{jr}) \quad \text{for} \quad j = Y, Z, {}^{0,2}X, \ldots, {}^{3,5}X$$

where m_p is the mass of p and δm_{jr} is the mass difference between ion type j and the B ion for residue r. These mass differences can be pre-computed and stored in a file. For any linkage b, p is called an i-support peak of PRF (m_{ir}, r, b). A tolerance of ϵ is applied for support peak computations. Using

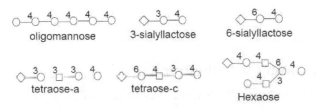

FIGURE 4.7: Glycan structures used to test the GLYCH algorithm. *Reused by permission of Oxford University Press.*

the fragmentation patterns that are supported by the peaks for a PRF, a score $s(F)$ can be computed for it as the number of supporting peaks N_f. Thus, the series of PRFs $F_1, F_2, \ldots, F_n$ that maximizes the sum $\sum s(F_i)$ should correspond to a single glycan structure.

In order to find the optimal series of PRFs, first, all PRFs are sorted according to their PRMs. For each PRF $F = (m, r, b)$, the score $V(m, r, b)$ is defined as the maximal total score of the series $(m_1, r_1, b_1), (m_2, r_2, b_2), \ldots, (m_j - m, r_j, b_j)$ where $m_1 \leq m_2 \leq \cdots \leq m_j$, and the structure rooted at r_j has total mass m. Thus the global optimal solution to the problem would be to solve $\max_{r \in \mathcal{R}} V(m_n, r, 1)$. This score $V(m, r, b)$ can be computed using dynamic programming over all PRFs in ascending order of their PRMs, as follows:

$$
V(m, r, b) = s(m, r, b) +
\min_{m_1 \leq m_2 \leq m}
\begin{cases}
0 & \text{if } m = m(r) \\
V(m_1, r_1, b_1) & \text{if } m = m(r) + m_1 \\
V(m_1, r_1, b_1) + V(m_2, r_2, b_2) & \text{if } m = m(r) + m_1 + m_2, \\
& \quad b_1 \neq b_2
\end{cases}
$$

where $m(r)$ is the mass of residue r. It is noted here that this algorithm assumes that each node of the glycan has at most two children. The resulting structure is obtained by backtracking through the computations.

This method was evaluated on test sets of MS/MS spectra of oligosaccharides, illustrated in Figure 4.7. These N-glycans were enzymatically released using ribonuclease B and permethylated before MS/MS analysis. The spectra were obtained by MALDI/TOF/TOF-MS. The peaks to analyze were selected using a sliding window of 20 peaks, where all peaks exceeding a threshold intensity determined by the average intensity minus three times the standard deviation were chosen. In the results, GLYCH was able to identify the real structure from among the best candidates for most of the structures. A re-evaluation procedure was used to improve this result by generating a theoretical spectrum from the oligosaccharides and ranking the best solutions according to the best match. In this case, the results improved for all the test structures, as shown in Table 4.8.

4.4   Data mining techniques

Data mining and machine learning algorithms for analyzing and classifying glycans have been developed as of late. These include kernel methods for classifying glycan structures and predicting potential glycan biomarkers, as well as probabilistic models for extracting glycan motifs. A primer on the

TABLE 4.8: GLYCH performance results. DP Rank refers to the initial ranking results, OS is the number of optimal solutions, and Re-evaluation rank is the ranking after the re-evaluation procedure.

Glycan	DP Rank	OS	Re-evaluation Rank	OS
Hexaose	1	369	1	26
3-Sialyllactose	1	31	1	2
6-Sialyllactose	1	35	1	2
Tetraose-a	1	61	1	3
Tetraose-c	1	58	1	2
Oligomannose	13	177	1	17

basics of these methods is provided in Appendix B. The latest methods in these areas will be described in this section.

4.4.1 Kernel methods

Kernel methods and support vector machines (SVMs) are now quite popular in the field of bioinformatics for classifying multi-dimensional data efficiently (Scholkopf and Smola (2002)). Users may refer to Appendix B.1 for a brief introduction to kernel methods and SVMs. For glycans, tree-based kernels have been developed that also take glycan-characteristic features into consideration. For example, the core structures are fairly consistent among glycans in the same class, but the structures at the leaves may actually be shared among glycans of different classes. It is presumed that these latter structures are key in recognition and signalling events.

Furthermore, the scoring system used in these kernels can be utilized in determining key structural features, in a method called feature extraction (Scholkopf et al. (2004)), that may potentially serve as glycan biomarkers. The first kernel method to attempt such glycan feature extraction is the layered trimer kernel (Hizukuri et al. (2005)). However, this method was developed especially for trimer structures, which was most specific for the leukemia data set that was being analyzed. In order to provide a more generalized kernel, the gram distribution kernel was developed. Around the same time, the multiple kernel was also developed. Each of these methods will be described in the following sections.

4.4.1.1 Layered trimer kernel

Glycan substructures at the non-reducing end are the structures that are believed to serve the more important roles in glycan function due to their flexibility, compared to the reducing end which tends to have less variety. These are the substructures that are recognized by other proteins and pathogens. It is also the site where glycosyltransferases catalyze glycosidic linkages.

TABLE 4.9:   Leukemia data set tested
by the layered trimer kernel. The control
data set was made up of glycans from other
blood components.

Blood component	Number of Structures
Leukemic cells	162
Erythrocytes	112
Serum	85
Plasma	73

Therefore, in the layered trimer kernel, a weighting scheme was developed which differentiated substructures based on the distance from the root, called "layer," which is defined as the number of glycosidic linkages between (the root of) the substructure and the root of the glycan. Furthermore, it was found that for the data set of glycans related to leukemic cells, trimer structures were the most effective in distinguishing between the different blood components being studied (leukemic cells, erythrocytes, serum and plasma). Trimer structures have also been implicated as the average size of structures recognized by glycosyltransferases. Thus the layered trimer kernel focused on substructures of size three. As a result, a feature vector for all possible trimer substructures at all possible layers was generated for the target glycan data set. These layered trimers, hereafter called features, were indexed such that they could be identified by numbers, corresponding to vector position. Each glycan was then assigned its own unique binary vector corresonding to whether or not it contained the feature at the given feature index. In the kernel calculation, given the feature vectors for two glycans X and Y, their inner product was calculated as $\sum_k w_k x_k y_k$, where k is a feature. The weighting parameter w_k was set to 1 when the layer of feature k was 1 in either of the two glycans. Otherwise, $w_k = 1 - \exp^{-\alpha h}$, where α was a positive constant to weight h, the minimum of the layers of the feature in glycans X and Y. That is, if the feature was in different layers in structures X and Y, the layer closer to the root was taken for the value of h. This would put less weight on structures near the root, emphasizing structures at the non-reducing end.

As an example, for the glycan structure in Figure 4.1, the layered trimers would be those displayed in Table 4.10. Note that the same structures appear in different layers. This allows the kernel to distinguish between structures at different layers.

Using this kernel, the leukemia-specific glycans in KEGG GLYCAN were trained against other glycans found specifically in other blood components, including erythrocyte, plasma and serum. Since kernel training performance is better with similarly-sized positive and negative data sets, the leukemia data set was compared against a randomly selected set of structures from the other blood components.

In the feature selection step, a scoring function was developed to score each

TABLE 4.10: Layered trimers for the glycan in
Figure 4.1. Note that the same structure appears multiple
times in different layers, thus allowing the kernel to distinguish
between them based on layer.

Layer	Structures
1	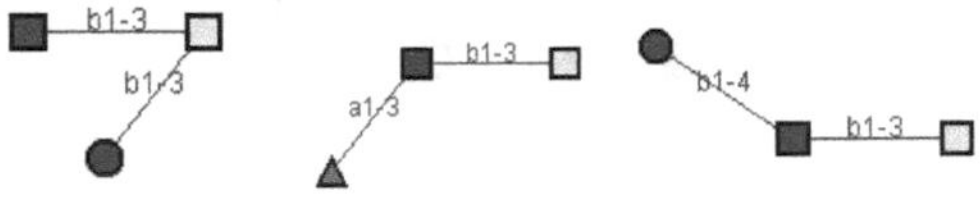
2	
3	
4	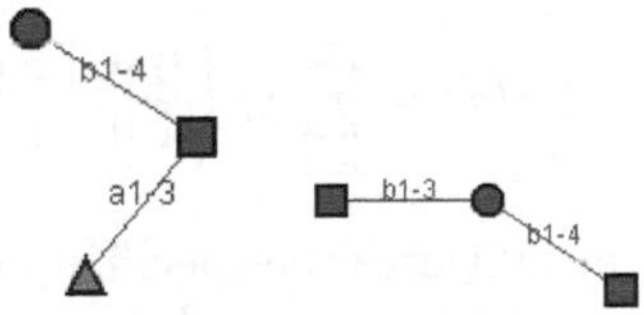
5	
6	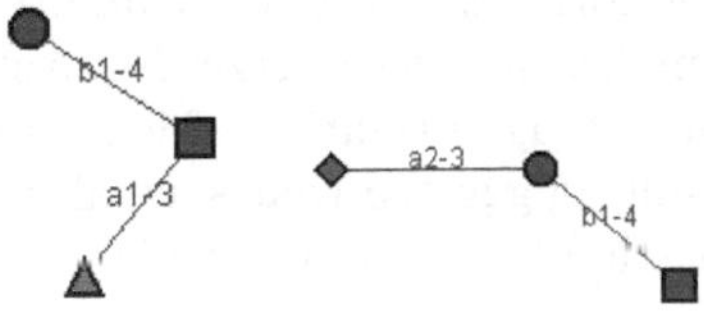

TABLE 4.11: Leukemic cell-specific glycan structures extracted by the layered trimer kernel, along with their specificity scores.

Feature	Layer	Specificity score
◆—a2-3—○—b1-4—■	5	161.2
○—b1-4—■—b1-2—○	4	159.6
◆—a2-6—○—b1-4—■	5	148.8
■—b1-2—○—a1-3—○	3	78.7
■—b1-2—○—a1-6—○	3	77.6

feature based on the trained kernel. The training process produces a discriminant score y for each target glycan X. Using these scores, the specificity z of each feature x can be computed by the following formula:

$$z(x) = \sum_{i=1}^{m} y_i \begin{cases} 0 \text{ if } & x \notin X_i \\ 1 \text{ if } & x \in X_i \end{cases}$$

This "specificity score" indicates the specificity of the feature given the target group of glycans compared to a control data set of glycans. The higher the score, the greater the specificity.

Given the dataset of glycans in various blood components listed in Table 4.9, the most leukemia-specific glycan structures turned out to be those that had been confirmed previously. These features are listed in Table 4.11.

The three highest-scoring features included sialic acid, which is known to appear in many tumor cells (Kannagi et al. (1986)). Furthermore, a cell agglutination assay was performed, whereby the addition of a structure corresponding to the top-scoring feature (specifically NeuAc(a2-3)-*N*-acetyl-lactosamine) indeed inhibited the cytoagglutination of leukemic cells from both T- and B-cell lines. These results suggested that several types of leukemic cells contain the same characteristic glycan motifs, which were successfully extracted by this kernel method.

4.4.1.2 Gram distribution kernel

The gram distribution kernel (Kuboyama et al. (2006)) used the concept of q-grams, which are defined as all possible paths of size q in a tree. A *path* corresponds to a chain; a linear structure having exactly two endpoints in the tree. Thus, traversing a path from one end, one may only travel in one direction in order to reach the other end. Figure 4.8 is an example of different q-grams for the given tree X. This idea came from the concept of a spectrum kernel for strings, containing all possible substrings of varying sizes, previously used for protein sequences (Leslie et al. (2009)).

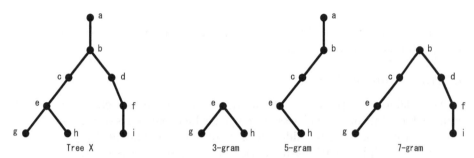

FIGURE 4.8: Example of the different q-grams of tree X.

In the q-gram kernel, the feature vector of each glycan structure is composed of the q-gram distribution of the structure for values of q within a certain range. Then the kernel function computes the dot product between two feature vectors to obtain the similarity score between the two glycan structures. The same feature extraction method as for the layered trimer kernel was used to extract specific glycan features for the target glycans.

Two test datasets were used to verify the performance of the q-gram distribution kernel. The first was the leukemia data set as tested previously, and the second was a data set for cystic fibrosis. The datasets tested are listed in Table 4.12. The number of structures used in this dataset is greater than that of the layered trimer kernel due to different data processing techniques used to extract the structures. However, it is assumed that this increase in data should not affect the classification results. In fact, it is desirable to have more, than less, data to train.

As a result, the performance of the q-gram distribution kernel was compared against the layered trimer kernel, and it was found that the performance was surprisingly comparable for the leukemia data set. In contrast, the q-gram distribution kernel outperformed the layered trimer kernel on the cystic fibrosis data set. A closer look at the extracted features showed that the most specific features were monomers and dimers, which could not be captured by the layered trimer kernel. In fact, it has been long known that the sulfated

TABLE 4.12:   Leukemia and cystic
fibrosis data sets tested by the q-gram
distribution kernel.

Glycan category	Number of Structures
Leukemic cells	191
Erythrocytes	274
Serum	202
Plasma	144
Cystic fibrosis	53
Respiratory mucin	123
Bronchial mucin	110

structures extracted by the q-gram kernel are specifically overexpressed in
cystic fibrosis (Xia et al. (2005)).

4.4.1.3   Multiple kernel

As a different approach to the q-gram distribution kernel, a new kernel
named the Multiple kernel was developed which defined several kernels for
glycans, each focusing on different sizes of subtrees and utilizing the layer
concept. Multiple kernel learning (Bach et al. (2005)) was used to compute
an optimal weighting of each kernel for each data set, and the performance was
evaluated by computing the area under the ROC curve (AUC), comparing the
number of true positives as a function of the false positives. Finally, feature
extraction methods were used to evaluate the features learned (Yamanishi
et al. (2007)).

For any particular kernel, the kernel function for two glycans t_1 and t_2 can
be expressed by the following function:

$$k(t_1, t_2) = \sum_{s_1 \in subtrees(t_1)} \sum_{s_2 \in subtrees(t_2)} q(s_1, s_2)$$

where $subtrees(t_1)$ are the subtrees of t_1 and $q(s_1, s_2)$ is the local kernel
function for the two subtrees s_1 and s_2. Note here that $subtrees(t)$ can also
represent any clearly defined set of substructures of t such as q-grams.

The local kernel functions that were evaluated include the following:

1. $q^0(s_1, s_2) = \delta(s_1 = s_2)$: This function evaluates to one (1) if $s_1 = s_2$ and
   zero (0) otherwise. The equality between trees is defined as the equality
   of the structure and all node and edge labels. Consequently, this kernel
   simply counts the number of common subtrees.

2. $q^N(s_1, s_2) = \delta(s_1 = s_2)\delta(n(s_1) = N)\delta(n(s_2) = N)$: Here, $n(s)$ denotes
   the number of nodes in subtree s, and N is a parameter for the number
   of nodes. This kernel thus counts the number of common subtrees of a
   particular size.

TABLE 4.13: Co-rooted trimers for the glycan in Figure 4.1, which consist of only those subtrees in which all siblings appear.

Layer	Structures
1	
2	
4	
6	

3. $q^D(s_1, s_2) = \max(D + 1 - |d(s_1) - d(s_2)|, 0)\delta(s_1 = s_2)$: Here, $d(s_i)$ is the layer of the root of s_i in tree t_i, and D is the maximum allowed difference in layers. When $D = 0$, only subtrees at the same layers are matched. It is noted here that the q^0 kernel can be considered as the limit of q^D as D approaches infinity.

4. Any product of q^N and q^D: A variety of common subtrees of size N and different layers can be counted.

As a variation to the set of all subtrees of any given tree, the concept of *co-rooted* subtrees was also considered. These are subtrees which contain all siblings; that is, either all siblings are included, or none at all. Table 4.13 displays the co-rooted trimers of the glycan in Figure 4.1.

In order to evaluate the multiple kernel, a preliminary evaluation of the local kernels was first performed on the same leukemia data set as described

TABLE 4.14:   AUC
performance of local kernel
functions used in training the
multiple kernel tested on the
leukemia dataset.

D	N	Co-rooted	All
0	1	92.0±0.6	89.7±1.0
0	3	91.1±0.4	93.3±0.2
0	all	92.5±0.2	91.5±0.6
2	1	91.7±0.1	89.9±0.5
2	3	93.3±0.3	94.6±0.2
2	all	93.2±0.2	92.1±0.5
none	1	91.5±0.2	91.0±0.6
none	3	92.9±0.3	94.0±0.2
none	all	92.2±0.1	92.7±0.0

in Section 4.4.1.1. In total, 266 local kernels were evaluated, for 19 values of N ranging between $1, \ldots, 18$ and *all* and seven values of D ranging from $0, \ldots, 5$ and *none*. Also considering the two datasets of all subtrees and all co-rooted subtrees, a total of $2 \times 7 = 266$ local kernels were defined. The AUC scores as a result of testing on this dataset using various values for $D(= 0, 2, none)$ and $N(= 1, 3, all)$ are listed in Table 4.14. As is evident from this table, the performance was generally high for all kernels, and there was not a significant difference in performance between the kernels using all subtrees and those using co-rooted subtrees. However, as expected, the trimer kernels achieved the highest performance among those with the same values for D.

The final step of extracting the leukemia cell-specific features was performed for each local kernel, resulting in the list of structures in Table 4.15. The highest scoring feature extracted by the multiple kernel is the same top-scoring feature extracted by the layered trimer kernel. However, there are also differently sized features appearing among the top six features. In particular, sialic acid is pronouncedly high as the second top feature, which coincides with the well-known fact that sialic acid-containing structures are often linked with cancer cells (Kannagi et al. (1986); Schauer (2000)).

Finally, the multiple kernel was used to classify the leukemia data set. The multiple kernel finds an optimal weighting of each local kernel for each data set. As a result, it achieved an AUC of 96.0±1.5, which was better than any of the individual kernels alone.

4.4.2   Frequent subtree mining

As was described in the previous sections, kernel methods provide a way to classify glycan structures, simultaneously extracting the most key features of the structures that distinguish between the two classes. This is performed by

TABLE 4.15: Leukemic cell-specific glycan structures extracted by the multiple kernel, along with their specificity scores.

Feature	Layer	Specificity score
◆ —a2-3— ○ —b1-4— ■	5	0.552
◆	7	0.502
○—b1-4—■ b1-6 ○—b1-4—■—b1-2—○—b1-6—○	3	0.467
○—b1-4—■ b1-6 ○—b1-4—■—b1-2—○—b1-6—○—b1-4—■	2	0.467
○—b1-4—■ b1-6 ○—b1-4—■—b1-2—○	4	0.467
◆ —a2-6— ○ —b1-4— ■	5	0.463

breaking down the glycan structures into smaller parts, from which the most important ones are selected using a mathematical formula.

One issue with glycan kernel methods is that the features vectors may tend to become quite large. The q-gram kernel broke down glycan structures into paths for values of $q = \{2, \ldots, 9\}$. However, q-grams are only a subset of subtrees, which include branched structures that are not considered paths. If a feature vector of all subtrees of a glycan were to be generated, this would require much computational space.

Thus a method for extracting those frequent subtrees from within a set of trees was developed and applied for glycan structures (Hashimoto et al. (2008)). By extracting only the most commonly occurring subtrees, the list of candidate substructures as biomarkers would become much more manageable.

This method takes as input a set of trees $\mathbf{T} = \{T_1, T_2, \ldots, T_n\}$, which are composed of the set of subtrees $\{t_1, t_2, \ldots, t_m\}$. The concept of *support* is then defined, where the support $S(t_i)$ of a subtree t_i is defined as the number of input trees in $\mathbf{T}$ that contain t_i. Thus one may define a *frequent subtree* as a subtree whose support is at least some value *minsup* (for minimum support). Applied to glycans, however, due to their high redundancy, a large number of frequent subtrees may result. To overcome this problem, another concept called *closed frequent subtrees* was defined as a frequent subtree t where no supertree containing t and having the same support exists. Figure 4.9 illustrates the concept of closed frequent subtrees. In the figure, the subtrees that are crossed out are not considered closed frequent subtrees because the supertrees A, B and C containing them also have the same support.

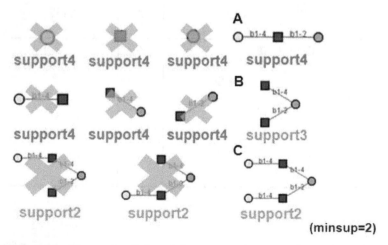

FIGURE 4.9: Example of closed frequent subtrees, where the subtrees A, B, and C are closed because no supertree containing them and having the same support exists.

TABLE 4.16: Classification performance of the α-closed frequent subtree mining method vs. the convolution kernel, the co-rooted subtree kernel, and the layered trimer kernel.

Method	AUC	Accuracy
α-closed frequent subtree	0.942	0.869
convolution kernel	0.934	0.857
co-rooted subtree kernel	0.916	0.843
layered trimer kernel	0.904	0.825

Furthermore, the concept of *maximal frequent subtrees* was also defined as a frequent subtree t having no frequent supertree containing t. That is, among the set of all frequent subtrees, say $\mathbf{T}$, any subtree not having any supertrees in $\mathbf{T}$ is considered a maximal frequent subtree. Thus, among the closed frequent subtrees in Figure 4.9, only subtree C would be considered a maximal frequent subtree since it contains the subtrees A and B.

In analyzing the closed frequent subtrees of the structures in the KEGG GLYCAN database, it was still found that a large number of subtrees resulted. In contrast, in terms of maximal frequent subtrees, only few would remain. Thus the concept of an α-closed frequent subtree was defined such that a new parameter α could serve to adjust the number of frequent subtrees extracted from a data set of trees. An α-closed frequent subtree is defined as a frequent subtree t satisfying the following equation: $support(t') \geq (\alpha \times support(t), minsup)$ for any supertree t' of t. Here, the value for α takes on a value between zero and one.

For verification, 10-fold cross validation was performed 10 times on a set of 485 O-glycans from the KEGG GLYCAN database. A data set of negative samples were generated for each of the ten runs by taking the positive data set and replacing the parent-child pairs based on the entire distribution of parent-child pairs from the positive data set. In training, the positive data was used for mining frequent subtrees, and both positive and negative data sets were used for hypothesis testing. This method was compared against other kernel methods called the convolution kernel (Kashima and Koyanagi (2002)), the co-rooted subtree kernel (Shawe-Taylor and Cristianini (2004)) and the Layered Trimer Kernel. The convolution kernel enumerates all possible subtrees of two trees and counts the common subtrees between them. The co-rooted and Layered Trimer Kernels were described previously. The classification performance results for $minsup = 2$ and $\alpha = 0.95$, which obtained the highest performance, are listed in Table 4.16.

Overall, the α-closed frequent subtree mining method outperformed all other kernel methods, indicating that for classification based on subtrees, frequent subtree mining methods may suffice. However, further analysis of biological function is still an open issue.

4.4.3   Probabilistic models

The development of probabilistic models for glycans grew out of the fact that the recognition mechanism of glycans by proteins such as lectins may involve not only the non-reducing end monosaccharides but also other sugars (internal nodes) further along the chain. This may also be a reason for why the same sugar motifs are often recognized by a variety of different pathogens, as described earlier in Section 1.6.

Before proceeding to describe these models, some of the notations that will be used will be defined here. The trees involved are all labeled ordered trees, which will hereafter be referred to as *trees*. A tree $T_i = (V_i, E_i)$ is a tree in the set $\mathbf{T} = \{T_1, T_2, \ldots, T_{|\mathbf{T}|}\}$ whose set of nodes are $V_i = \{q_1^i, q_2^i, \ldots, q_{|V_i|}^i\}$ and edges are $E_i \subseteq V_i \times V_i$. Let q_1^i be the root of tree T_i. If understood from the context, nodes may not be notated with the tree index i (i.e., q_j) and they may also simply referred to by their indices where node j refers to q_j^i and p is its parent node. The nodes V_i are indexed in breadth-first order such that all children of a node are traversed before any of the grandchildren. Thus, for any node q_j^i in V_i, the immediate elder sibling is indexed at $j-1$ and the immediate younger sibling at $j+1$, notated as q_{j-} and q_{j+}, respectively. $t_i(j)$ is defined as the subtree of T_i whose root is q_j^i. $q_\leftarrow^i(p)$ is defined as the eldest child of node q_p^i. Conversely, the youngest child is $q_\rightarrow^i(p)$. Let $C(p) \subseteq \{1, 2, \ldots, |V_i|\}$ be the set of indices of the children of q_p^i, and $|C| = \max_{i,p} |C_i(p)|$. $Y_i(j)$ is the set of indices of all younger siblings of q_j^i. Additionally, $o_j^i \in \Sigma$ is the output label for node q_j^i, where $\Sigma = \{w_1, w_2, \ldots, w_{|\Sigma|}\}$ is the set of labels. To define the probabilistic model, the notations corresponding to those used in Appendix B.2 are used as well.

4.4.3.1   Probabilistic sibling-dependent tree Markov model

For an introduction to Markov models, hidden Markov models (HMMs) and hidden tree Markov models (HTMM), readers may refer to Appendix B.2. The first probabilistic model for glycans was the probabilistic sibling-dependent tree Markov model, or PSTMM (Aoki et al. (2004)). The reason this model was developed as opposed to using the HTMM model was due to the possible patterns of glycans that could be recognized, involving sibling-dependencies which could not be captured with HTMM. Although adding more dependencies would require more complicated and time-consuming algorithms for learning the model, sufficiently efficient algorithms within reasonable bounds were developed while achieving statistically significant results in predictive performance.

A schematic of PSTMM is provided in Figure 4.10a. The PSTMM for the tree at the far left consists of sibling dependencies and parent-child dependencies indicated by dotted arrows. Thus for any node, there are at most two dependencies; one to an older sibling and one to its parent. Only the "eldest children" have a single dependency to their parent. Each node also holds the

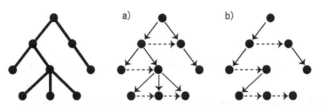

FIGURE 4.10: Schematics of a) PSTMM and b) OTMM corresponding to the tree on the far left. For both models, the sibling dependencies are indicated by the dotted arrows. The dependencies in PSTMM include all the parent-child dependencies as in HTMM, whereas in OTMM, the parent-child dependencies between only the eldest child and parent remain.

probability of outputting a label from the given alphabet for the model. In the case of glycans, this alphabet would be the set of monosaccharide names.

PSTMM contains three probability parameters, π, a and b, that must be learned for a given tree t. The initial state probability $\pi[s]$ for a given state s is defined as the probability that state s is the state of the root node of t. For a node q_n, whose elder sibling node is q_b and parent node is q_p, the state transition probability $a[\{s_p, s_b\}, s_n]$ is the conditional probability that state s_n is the state at q_n, given that s_p is the state at q_p and s_b is the state at q_b. For completeness, the state transition probability $a[\{s_p, -\}, s_n]$ is defined for $q_{\leftarrow}(p)$ which is the conditional probability that the state of $q_{\leftarrow}(p)$ is s_n and the state of q_p is s_p. Finally, the label output probability is $b[s_b, w_h] = p\{o_j^i = w_h | q_j^i = s_b, \lambda\}$, which is the conditional probability that the output label is w_h given that s_b is the state at node q_j in tree t_i. Note that $\Sigma_l \pi[l] = 1$, $\Sigma_m a[\{s_q, s_l\}, s_m] = 1$, and $\Sigma_h b[l, w_h] = 1$.

4.4.3.1.1 Probability evaluation In order to compute the probability of a sequence (or, in the case of glycans, a tree structure) of observations given our PSTMM, auxiliary variables can be used, as was done for HMMs. The upward probability $U_p^i(s_q)$ is the probability that all labels of subtree $t_i(p)$ are generated and that the state of node q_p^i is s_q. The forward probability $F_j^i(s_q, s_l)$ is the probability that for node q_j^i, all labels of the subtrees of all elder siblings are generated and that the states of node q_j^i and its parent q_p^i are s_l and s_q, respectively. For completeness, the backward probability $B_j^i(s_q, s_m)$ is also defined, referring to the probability that for node q_j^i, all labels of the subtrees of all younger siblings and itself are generated and that the states of node q_j^i and its parent are s_m and s_q, respectively.

These auxiliary probabilities can be mathematically defined as follows.

$$U_p^i(s_q) = \begin{cases} b[s_q, w_p^i] & \text{if } C(q_p^i) = \emptyset \\ b[s_q, w_p^i] \sum_m^{|S|} F_j^i(s_q, s_m) B_j^i(s_q, s_m) & \text{otherwise} \end{cases}$$

where $q_j^i \in C(q_p^i)$.

$$F_j^i(s_q, s_l) = \begin{cases} a[\{s_q, -\}, s_l] & \text{if } q_j^i = q_{\leftarrow}^i(p) \\ \sum_m^{|S|} Fj{-}^i(s_q, s_m)U_{j-}^i(s_m)a[\{s_q, s_m\}, s_l] & \text{otherwise} \end{cases}$$

$$B_j^i(s_q, s_m) = \begin{cases} U_j^i(s_m) & \text{if } q_j^i = q_{\rightarrow}^i(p) \\ U_j^i(o_m)\sum_l^{|S|} a[\{s_q, s_m\}, s_l]B_{j+}^i(s_q, s_l) & \text{otherwise} \end{cases}$$

Thus by estimating the parameters using these auxiliary probabilities (as described in the next section), the likelihood of a given tree structure for a PSTMM can be obtained using $U_1^i(s_l)$, the upward probability at the root of the tree t_i, as follows:

$$L(\mathbf{T}) = \prod_{i=1}^{|\mathbf{T}|} L(T_i) = \prod_{i=1}^{|\mathbf{T}|} \sum_{m=1}^{|S|} \pi[s_m]U_1^i(s_m)$$

4.4.3.1.2 Parameter estimation The parameters are estimated using the EM algorithm (Dempster et al. (1977)), which uses a fourth auxiliary probability called the downward probability $D_j^i(s_l)$, defined as the probability that all labels of tree t_i except its subtree $t_i(q_j)$ are generated and that the state of node q_j^i is s_l. This parameter can be computed at a node by using the downward probability at its parent and the forward and backward probabilities of its elder and younger siblings, respectively, as follows.

$$D_j^i(s_l) = \begin{cases} \pi[s_l] & \text{if } j = 1 \\ \sum_n^{|S|} D_p^i(s_n)b[s_n, w_p^i]F_j^i(s_n, s_l) & \text{if } q_j^i = q_{\rightarrow}^i(p) \\ \sum_n^{|S|} D_p^i(s_n)b[s_n, w_p^i]F_{j-}^i(s_n, s_l)\sum_m^{|S|} a[\{s_n, s_l\}, s_m]B_{j+}^i(s_n, s_m) & \text{o.w.} \end{cases}$$

These four probabilities can be estimated using an algorithm that traverses the nodes of the tree in linear time. The pseudocode for this algorithm is given in Figure 4.11. There are four basic procedures to be called, as shown on lines 2 through 5: The procedure `estimateParams` performs a depth-first traversal of the tree from the root to the leaves. Once the leaves are reached, the upward parameter is computed for each node (line 12). The details of this procedure shows that after computing the upward parameter, the procedure is called recursively for each younger sibling, thus traversing all the leaves. Then line 13 calls the main procedure to compute the forward and backward probabilities, similar to the *forward-backward* algorithm for HMM (see Appendix B.2.2). However, since these parameters depend on two states, two nested `for` loops must be called (lines 25 through 29) to compute the forward parameter. If the current node has a younger sibling, this procedure is recursively called until the forward parameter for all siblings have been computed.

Conversely, once the forward parameters have been computed, the backward probability is recursively computed going backward up to the eldest sibling via the `estimateBackwardForward` procedure. This procedure is then called while returning back from the depth-first traversal on lines 12 and 13. Finally, the downward parameter is computed from the root down to the children in breadth-first order on lines 49 through 55. Thus by traversing the tree from the root to the leaves, computing the upward, forward and backward parameters returning to the root, and then computing the downward parameters returning down, all parameters can be computed in $O(|V|)$ time.

Using these four auxiliary probability parameters, the EM algorithm can be used to compute expectation values for the probability parameters for a tree T_i. For the state transition parameter $a(\{s_l, s_m\}, s_n)$, the expectation value $\gamma_i(\{s_l, s_m\}, s_n)$, which is the expectation that the state of a node is s_n and that the states of its parent and immediately elder sibling are s_l and s_m, respectively, can be computed. For states having no elder sibling, the corresponding expectation value is computed as $\gamma_i(\{s_l, -\}, s_n)$. For the label output probability $b[s_m, w_h]$, the expectation value $\delta_i(s_m, w_h)$ is computed for the expectation that the state of a node is s_m and the output label is w_h. Finally, for the initial state probability $\pi[s_m]$, $\eta_i(s_m)$ can be computed, which is the expectation that the initial state of node q_1^i is s_m. The equations for these expectation values are as follows.

$$
\gamma_i(\{s_l, s_m\}, s_n) =
$$
$$
\frac{1}{L(T_i)} \sum_{q_p^i : C(p) \neq \emptyset} D_p^i(s_l) b[s_l, w_p^i] \sum_{j \in Y(p)} F_{j+}^i(s_l, s_m) U_{j+}^i(s_m) a[\{s_l, s_m\}, s_n] B_j^i(s_l, s_n),
$$
$$
\gamma_i(\{s_l, -\}, s_n) = \frac{1}{L(T_i)} \sum_{q_p^i : C(p) \neq \emptyset} D_p^i(s_l) b[s_l, w_p^i] a[\{s_l, -\}, s_n] B_k^i(s_l, s_n)
$$

where $q_k^i = q_{\rightarrow}(p)$ and $Y(p)$ is all the children of p except $q_{\leftarrow}(p)$.

$$
\delta_i(s_m, w_h) = \frac{1}{L(T_i)} \sum_{q_j^i : o_j^i = w_h} D_j^i(s_m) \, U_j^i(s_m),
$$
$$
\eta_i(s_m) = \frac{1}{L(T_i)} \pi[s_m] \, U_1^i(s_m).
$$

```
1  procedure estimatePSTMMparams() {
2      estimateParams(q₁);
3      estimateUpward(q₁);
4      estimateForwardBackward(q₁);
5      estimateDownward(q₁);
6  }
7  procedure estimateParams(qⱼ) {
8      /* for all children of qⱼ, traverse nodes from oldest to youngest */
9      for each c ∈ C(qⱼ) do
10         estimateParams(c)
11     /* traverse from oldest to youngest child */
12     estimateUpward(q←(qⱼ));
13     estimateForwardBackward(q←(qⱼ));
14 }
15 procedure estimateUpward(qⱼ) {
16     for each s_q ∈ S do
17         compute Uⱼ(s_q);
18     end
19     /* compute U of next younger sibling */
20     if qⱼ has younger sibling do
21         estimateUpward(q_{j+});
22     end
23 }
24 procedure estimateForwardBackward(qⱼ) {
25     for each s_l ∈ S do
26         for each s_m ∈ S do
27             compute Fⱼ(s_l, s_m);
28         end
29     end
30     if qⱼ has younger sibling do
31         /* compute F of immediately younger sibling */
32         estimateForwardBackward(q_{j+});
33     else /* compute B of immediately elder sibling */
34         estimateBackwardForward(q_{j-});
35     end
36 }
37 procedure estimateBackwardForward(qⱼ) {
38     for each s_l ∈ S do
39         for each s_m ∈ S do
40             compute Bⱼ(s_l, s_m);
41         end
42     end
43     /* compute B of immediately elder sibling */
44     if qⱼ has elder sibling do
45         estimateBackwardForward(q_{j-});
46     end
47 }
48 procedure estimateDownward(qⱼ) {
49     for each s_l ∈ S do
50         compute Dⱼ(s_l);
51     end
52     /* for all children c of qⱼ */
53     for each c ∈ C(qⱼ) do
54         estimateDownward(c)
55     end
56 }
```

FIGURE 4.11: Pseudocode for calculating F, B, U and D in PSTMM

Using γ_i, δ_i and η_i, the PSTMM probabilities can be updated as follows:

$$\hat{a}[\{s_l, -\}, s_n] = \frac{\sum_i \gamma_i(\{s_l, -\}, s_n)}{\sum_i \sum_m \gamma_i(\{s_l, -\}, s_m)},$$

$$\hat{a}[\{s_l, s_m\}, s_n] = \frac{\sum_i \gamma_i(\{s_l, s_m\}, s_n)}{\sum_i \sum_k \gamma_i(\{s_l, s_m\}, s_k)},$$

$$\hat{b}[s_m, w_h] = \frac{\sum_i \delta_i(s_m, w_h)}{\sum_i \sum_j \delta_i(s_m, w_j)},$$

$$\hat{\pi}[s_m] = \frac{\sum_i \eta_i(s_m)}{\sum_i \sum_l \eta_i(s_l)}.$$

After updating these probability parameters, the likelihood for the input set of trees is computed and compared with the likelihood before updating. If the difference between the likelihood is small enough (i.e., it has converged), then the likelihood is maximized and the algorithm ends.

4.4.3.1.3 Optimal state sequence

Once the likelihood is maximized, the states and transition paths that contributed most to the likelihood can be computed. This can be done by first computing the maximum values for the parameters B and U as follows:

$$\phi_U(s_l, q_p^i) = \begin{cases} b[s_l, w_p^i] & \text{if } C_i(q_p^i) = \emptyset \\ \max_m b[s_l, w_p^i] a[\{s_l, -\}, s_m] \phi_B(s_l, s_m, q_{\leftarrow}^i(p)) & \text{otherwise} \end{cases}$$

$$\phi_B(s_l, s_m, q_j^i) = \begin{cases} \phi_U(s_m, q_j^i) & \text{if } q_j^i = q_{\rightarrow}^i(p) \\ \max_n \phi_U(s_m, q_j^i) a[\{s_l, s_m\}, s_n] \phi_B(s_l, s_n, q_{j+}) & \text{otherwise} \end{cases}$$

Here, $\phi_U(s_l, q_p^i)$ computes for node q_p^i, the maximum probability that all labels of the subtree $t_i(p)$ are generated and that it is in state s_l. Correspondingly, $\phi_B(s_l, s_m, q_j^i)$ computes for node q_j^i the maximum probability that it is in state s_m, that its parent q_p is in state s_l, and that all labels of the subtrees rooted at all its younger siblings have been generated.

Finally, knowing the maximum probabilities, the states producing these probabilities can be retrieved by computing τ_U and τ_B using $\arg\max^3$ of ϕ_U and ϕ_B, respectively.

$$\tau_U(s_l, q^i_p) = \begin{cases} 0 & \text{if } C_i(q^i_p) = \emptyset \\ \arg\max_m b[s_l, w^i_p]a[\{s_l, -\}, s_j]\phi_B(s_l, s_m, q_\leftarrow(p)) & \text{otherwise} \end{cases}$$

where s_j is the state of $q_\leftarrow(q^i_p)$.

$$\tau_B(s_l, s_m, q^i_j) = \begin{cases} 0 & \text{if } q^i_j = q_\rightarrow(p) \\ \arg\max_n \phi_U(s_m, q^i_j)a[\{s_l, s_m\}, s_n]\phi_B(s_l, s_n, q_{j+1}) & \text{otherwise} \end{cases}$$

By computing $q^*_j = \tau_U(q^*_p)$ for $q_j = q_\leftarrow(p)$ and $q^*_j = \tau_B(q^*_{j+})$ for all other nodes, the set of states $\{q^*_1, \ldots, q^*_{|V_i|}\}$, which is the most likely state transition path for the given tree T_i, can be retrieved.

4.4.3.1.4 Experimental results of validation on glycan data
The PSTMM model just described was next validated by evaluating its performance on both experimental and biological data. In preparing actual glycan structures for validation, the node and edge information for each structure from the KEGG GLYCAN database was read, ordering the children of each node according to the carbon number to which they were attached to their parent. Thus, each node j corresponded to a monosaccharide, and each immediately younger sibling $j-$ corresponded to the monosaccharide attached with the next higher carbon number. For example, Figure 4.12 illustrates how two GlcNAc (■) nodes attached to mannose (○) are ordered; for G04023, the lower child is attached to C2 while the upper child is attached to C6.

For validation, the performance of PSTMM was compared against two other simpler models called Label Pair Model (LPM) and Mixture of LPMs (MLPM). LPM simply counts the frequency of occurrence of each parent-child pair. MLPM is a probabilistic model having the following parameters: $z[c, w_h, w_{h'}]$ ($\sum_h z[c, w_h, w_{h'}] = 1$ for each pair (c, h')) and $\pi[c, w_h]$ ($\sum_h \pi[c, w_h] = 1$ for each c). For a component c, $z[c, w_h, w_{h'}]$ ($= P(o_j = w_h | o_p = w_{h'}, c)$) is the conditional probability that label w_h is outputted at a node given that $w_{h'}$ is outputted at its parent node, and $\pi[c, w_h]$ ($= P(o_1 = w_h | c)$) is the probability that the root label is w_h.

LPM is simply an MLPM containing just one component, so no iteration of the EM algorithm is applied to LPM; that is, its parameters are calculated exactly once. The estimation procedure for MLPM has been introduced in a review on mixture models by McLachlan and Peel (2000).

---

[3]The argument of the maximum, referring to the value of the given argument for which the value of the given expression obtains its maximum value. Thus, $\arg\max_x f(x)$ refers to the value of x for which $f(x)$ is maximum.

TABLE 4.17: Glycan experiment results for (a)
N-Glycans, (b) *O*-Glycans, (c) Glycosaminoglycans,
and (d) Sphingolipids using five-fold cross validation.

		PSTMM	MLPM	LPM
	AUC	0.92	0.678 (**28.5**)	0.551 (**45.5**)
a	Accuracy	0.855	0.645 (**22.5**)	0.554 (**33.6**)
	Precision	0.956	0.668 (**29.2**)	0.557 (**39.8**)
	AUC	0.801	0.649 (**11.4**)	0.549 (**24.4**)
b	Accuracy	0.753	0.638 (**10.5**)	0.571 (**19.2**)
	Precision	0.841	0.627 (**13.5**)	0.550 (**20.1**)
	AUC	0.919	0.696 (**10.3**)	0.487 (**24.0**)
c	Accuracy	0.864	0.672 (**11.3**)	0.537 (**23.8**)
	Precision	0.963	0.724 (**9.6**)	0.489 (**24.4**)
	AUC	0.883	0.651 (**14.3**)	0.590 (**28.0**)
d	Accuracy	0.831	0.650 (**12.8**)	0.617 (**19.0**)
	Precision	0.929	0.641 (**14.5**)	0.613 (**19.3**)

The likelihood L for a given set of trees is computed by MLPM as $L = \prod_i^{|T|} \sum_c p(c)\pi[c, w_1^i] \prod_j a[c, w_j^i, w_p^i]$.

It is noted that for capturing patterns based on multiple parent-child relationships in a given set a trees, MLPM has the same representational power as that of PSTMM. Therefore, using MLPM in these experiments to compare with the performance of PSTMM suffices to prove its performance advantage.

Four glycan classes from KEGG GLYCAN were selected as the datasets for validation: *N*-Glycans, *O*-Glycans, Glycosaminoglycans, and Sphingolipids. The other classes were disregarded due to either an insufficient number of glycans or an insufficient average glycan size (i.e., number of nodes in each tree). The structures within each of these four classes were purged of any trees that did not have siblings such that only those structures that contained at least one sibling pair would be analyzed.

Five-fold cross-validation for the glycan structures within each class was performed, where each data set was divided into five subsets containing randomly selected tree structures from that class. Each subset was tested in one round for a total of five rounds. For each test round, 50 randomly selected structures from each of the non-test sets were trained for a total of 200 training structures, and all of the structures in the test set were tested for that round. A corresponding negative example test set was also tested. This test set consisted of a set of trees whose tree size (i.e., number of nodes) and parent-child label pair distribution was equivalent to that of the positive test set. The negative test set was thus created so that the simpler models would not be able to easily distinguish between the positive and negative test sets.

The performance of PSTMM was compared against the two simpler models using the following parameters[4]: $|\mathbf{T}| = 200$ for training, $|S| = 10$, $|\Sigma| = 19$, $|C| = 5$ and number of components in MLPM = 10. Note that as the trees in each data set varied in tree size, the likelihood calculation for each tree needed to be corrected accordingly. Therefore, each probability parameter value was multiplied by its size $(a[\{s_q, s_m\}, s_l] * |S|)$. These corrected parameter values (or scores) were used to calculate the likelihood of each tree. Finally, this entire experiment was repeated five times.

The results were averaged over the 25 (5 × 5) runs, which is listed in Table 4.17, computing the AUC, prediction accuracy, and precision (at sensitivity of 0.3), for the three methods tested on four classes of glycans. AUC refers to the area under the ROC (Receiver Operator Characteristic) curve (Hand and Till (2001); Hanley and McNeil (1982)) and can be computed by first sorting the examples by their likelihoods and then by using the following equation:

$$
\text{AUC} = \frac{R_n - \frac{n_n \cdot (n_n + 1)}{2}}{n_n \cdot n_p},
$$

where n_n (n_p) is the number of negative (positive) samples and R_n is the sum of the ranks of the negative samples. Note that $n_n = n_p$ in the experiments.

An AUC value takes on a value between 1 and 0 (the higher, the better) and is defined as the false positive threshold at zero sensitivity, where the false positive threshold is based on the false positive rate, which is the proportion of the number of false positives to the total number of negative examples, and sensitivity is the proportion of the number of correctly predicted examples to the total number of positive examples. Prediction accuracy is defined as the threshold at which the positive and negative test scores are best discriminated, and precision is the proportion of correctly predicted examples to the number of examples predicted to be positive. For these experiments, a reasonable sensitivity value of 30% was selected.

Table 4.17 lists the t-values[5] in parentheses, indicating that PSTMM statistically outperforms both LPM and MLPM by a significant margin. The N-Glycan class had the best performance among all four classes, which may be because of its large dataset size. However, even with a small dataset size such as Sphingolipid, PSTMM had a considerable performance advantage. It is apparent that there indeed exist long-range dependencies across siblings

---

[4]For $|\mathbf{T}|$, 50 training trees were randomly selected from each of the four non-test sets. For this experiment, each state was allowed to transition to any other state. However, as the main patterns in each class are better understood, more limitations may be placed on the paths through which the states can transition. For $|\Sigma|$, a scan through all the structures in KEGG Glycan revealed 19 various monosaccharides, so $|\Sigma| = 19$.

[5]t-values indicate the significance of the difference between two sets of values; if the t-value is larger than a certain value, say 8.610, then it can be claimed that the performance advantage of PSTMM is statistically significant over the other models at confidence level 99.9%.

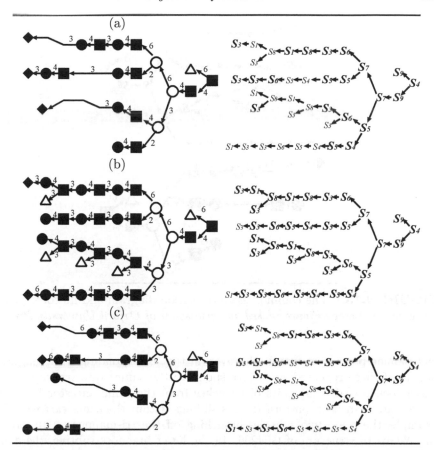

FIGURE 4.12: KEGG glycans: (a) G04023, (b) G04206, and (c) G03990. The top structures are the actual glycans, and the bottom structures are the most likely state transition diagrams. *Figure reused by permission of Oxford University Press.*

that could not be captured by any of the other methods. In considering such results, the increased time complexity is well worth the information gained from this model.

The most likely states learned from the datasets were analyzed in order to find the most likely state transitions. These would correspond to common patterns in the datasets and could be used to perform multiple tree structure alignment. Figure 4.12 illustrates three tree structures that PSTMM found to have similar patterns. The state transition model learned from these structures is given to the right of each glycan structure, with the state corresponding to each monosaccharide emphasized in bold. It was interesting to see that, for instance, in the largest of these three glycans, G04206, many repeated

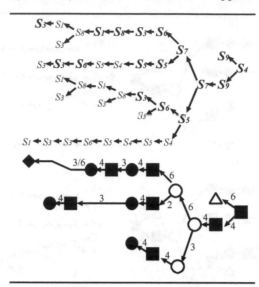

FIGURE 4.13:   The common state transition diagram and its corresponding glycan subtree. *Figure reused by permission of Oxford University Press.*

lactosamine pairs were found. However, in the corresponding state diagram,
not all of these repeated pairs correspond to the same state transitions. A
closer look reveals that pairs of branches from the same ancestor have the
same state transition pattern (across siblings within the same *subtree*). For
example, there are two subtrees branching off the tri-mannose ($\bigcirc$) core of
the N-Glycan structure of G04206. In the lower mannose subtree, there are
two subtrees, both of which contain lactosamine twice in sequence. However,
the upper branch of this sequence corresponds to a state transition path of
$S_6 - S_3 - S_8 - S_1$, while the lower branch corresponds to $S_5 - S_6 - S_3 - S_3$. In
the upper mannose subtree, the same pairs of monosaccharide branches and
the same two sets of state transition paths can be found. Furthermore, the
other two glycans, G03990 and G04023, also contain the same pattern near
their leaves. These different state transition patterns, despite representing the
exact same sugar chains, indicate possible functional roles corresponding to
the different states.

Figure 4.13 illustrates the full pattern that matched across these three glycans; the intersection of the three state transition diagrams is given along
with the corresponding glycan pattern fragment. The multiple tree alignment can be derived from these diagrams as each tree is aligned according
to the common state transition pattern found by PSTMM. However, it is
noted that the lowest branch from each of these glycans is *not aligned at all*
because although each of these lowest branches match in terms of pairs of
the lactosamine motifs, they correspond to different states. Therefore, they
are actually considered not to align with each other according to their sibling

relationships.

4.4.3.2 Ordered tree Markov model (OTMM)

The PSTMM model was further improved and shown to achieve similar predictive accuracy as the original model. The new model, called ordered tree Markov model, or OTMM (Hashimoto et al. (2006)), used the same parameters as PSTMM but with a new dependency model, as shown in Figure 4.10b. In OTMM, it was thought that since the eldest child has a single dependency on its parent, the younger siblings need only depend on their immediately elder siblings. Dependencies to the parent could be captured via the eldest sibling through the property of Markov models to capture long-range dependencies. Thus the complexity of the model was reduced to that of HTMM while still capturing the sibling relationships in the tree. That is, OTMM is a first-order Markov chain model, where each state only depends on one other state.

OTMM uses the same probability parameters as PSTMM, except the algorithms have been simplified. In particular, the state transition probability now only depends on the state for either the parent node or elder sibling node. So two different types of state transition probabilities are now defined:

$$a_p[s_l, s_m] = p\{q_j^i = s_m | q_p^i = s_l, \lambda\}$$

$$a_b[s_l, s_m] = p\{q_j^i = s_m | q_{j-}^i = s_l, \lambda\}$$

These correspond to the conditional probabilities that when a node is at state s_m its parent is s_q or its elder brother is s_l, respectively. The label output probability remains the same: $b[s_l, w_h] = p\{o_j^i = w_h | q_j^i = s_l, \lambda\}$. Note that $\sum_l \pi[s_l] = 1$, $\sum_m a_p[s_l, s_m] = 1$, $\sum_m a_b[s_l, s_m] = 1$, and $\sum_h b[s_l, w_h] = 1$.

To address the three problems of interest for a probabilistic model (described in Appendix B.2.1), the upward and backward probabilities for OTMM are again defined. The upward probability $U_p^i(s_l)$ is the probability that all labels of subtree $t_i(q_p)$ are output and that the state of node q_p is s_l. The backward probability $B_i(s_l, q_j)$ is the probability that for a node q_j, all labels of the subtrees for all younger siblings including node q_j are output, and that the state of q_j is s_l. These probabilities can be computed using a dynamic programming method that traverses the tree from the leaves up to the root.

$$U_p^i(s_l) = \begin{cases} b[s_l, w_p^i] & \text{if } C_i(q_p^i) = \emptyset \\ b[s_l, w_p^i] \sum_{m=1}^{|S|} a_p[s_l, s_m] B_k^i(s_m) & \text{otherwise} \end{cases}$$

where $q_k^i = q_{\leftarrow}(p)$.

$$B_j^i(s_m) = \begin{cases} U_j^i(s_m) & \text{if } q_j^i = q_{\rightarrow}^i(p) \\ U_j^i(s_m) \sum_{l=1}^{|S|} a_b[s_m, s_l] B_{j+}^i(s_l) & \text{otherwise} \end{cases}$$

As a result, the likelihood for a given tree T_i for this model would be computed by using the upward probability at the root of the tree, as follows:

$$L(T_i) = \sum_{m=1}^{|S|} \pi[s_m]U_1^i(s_m)$$

Consequently, the likelihood for the set of trees is defined as the product of the likelihood for each tree in the set:

$$L(\mathbf{T}) = \prod_{i=1}^{|\mathbf{T}|} L(T_i) = \prod_{i=1}^{|\mathbf{T}|} \sum_{m=1}^{|S|} \pi[s_m]U_1^i(s_m)$$

which provides the solution to the first problem of interest.

Next, to compute the parameters for the model, forward and downward probabilities also need to be defined. The forward probability $F_j^i(s_l)$ is the probability that all labels of tree T_i except for those of the subtrees rooted at q_j^i, all younger siblings of q_j^i are output, and the state of q_j^i is s_l. The downward probability $D_j^i(s_l)$ is the probability that all labels of tree T_i except for those of subtree $t_i(j)$ are output and that the state of q_j^i is s_l.

These two probabilities can be computed from the root going downward as follows:

$$F_j^i(s_l) = \begin{cases} \sum_n^{|S|} a_p[s_n, s_l]D_p^i(s_n)b[s_n, w_p^i] & \text{if } q_j^i = q_{\to}^i(p) \\ \sum_{m=1}^{|S|} a_b[s_m, s_l]F_{j-}^i(s_m)U_{j-}^i(s_m) & \text{otherwise} \end{cases}$$

$$D_j^i(s_l) = \begin{cases} \pi[s_l] & \text{if } j = 1 \\ F_j^i(s_l) & \text{if } q_j^i = q_{\to}^i(p) \\ F_j^i(s_l) \sum_{m=1}^{|S|} a_b[s_l, s_m]B_{j+}^i(s_m) & \text{otherwise} \end{cases}$$

The forward probability at the eldest sibling is computed using the downward probability of its parent, and each younger sibling uses the forward probability of its immediately elder sibling. The downward probability at a node is computed using its own forward probability and the backward probability of any younger siblings. Thus, the forward probability for a node must be computed before the downward probability is computed.

Using these four probabilities, the EM algorithm to estimate the optimum parameters can now be defined. The expectation values to compute are $\mu_i(a_p[s_n, s_l])$, $\mu_i(a_b[s_n, s_l])$, $\mu_i(b[s_m, w_h])$, and $\mu_i(\pi[s_m])$, as follows.

$$\mu_i(a_p[s_n, s_l]) = \frac{1}{L(T_i)} \sum_{c:C_i(p)\neq\emptyset} D_c^i(s_n)b[s_n, w_p^i]a_p[s_n, s_l]B_j^i(s_l),$$

where $q_j^i = q_\leftarrow^i(p)$.

$$\mu_i(a_b[s_n, s_l]) = \frac{1}{L(T_i)} \sum_{j:Y_i(j)\neq\emptyset} F_j^i(s_n)a_b[s_n, s_l]B_{j+}^i(s_l)U_j^i(s_n)$$

$$\mu_i(b[s_m, w_h]) = \frac{1}{L(T_i)} \sum_{k:o_k^i=w_h} D_k^i(s_m)U_k^i(s_m)$$

$$\mu_i(\pi[s_m]) = \frac{1}{L(T_i)}\pi[s_m]U_1^i(s_m)$$

Using these expectation values, the probability parameters can be updated in the following manner.

$$\bar{a}_p[s_n, s_l] = \frac{\sum_i \mu_i(a_p[s_n, s_l])}{\sum_i \sum_m \mu_i(a_p[s_n, s_m])}$$

$$\bar{a}_b[s_n, s_l] = \frac{\sum_i \mu_i(a_b[s_n, s_l])}{\sum_i \sum_m \mu_i(a_b[s_n, s_m])}$$

$$\bar{b}[s_m, w_h] = \frac{\sum_i \mu_i(b[s_m, w_h])}{\sum_i \sum_m \mu_i(b[s_m, w_m])}$$

$$\bar{\pi}[s_m] = \frac{\sum_i \mu_i(\pi[s_m])}{\sum_i \sum_k \mu_i(\pi[s_k])}$$

Finally, to address the most likely state sequence from the trained model, the upward and backward probabilities can be reformulated to find the maximum probabilities $\phi_U(s_n, q_p^i)$ and $\phi_B(s_m, q_j^i)$ as follows:

$$\phi_U(s_n, q_p^i) = \begin{cases} b[s_n, w_p^i] & \text{if} \quad C_i(p) = \emptyset \\ b[s_n, w_p^i] \max_{m=1}^{|S|} a_p[s_n, s_m]B_k^i(s_m) & \text{otherwise} \end{cases}$$

where $q_k^i = q_\leftarrow^i)(p)$.

$$\phi_B(s_m, q_j^i) = \begin{cases} U_j^i(s_m) & \text{if} \quad q_j^i = q_\rightarrow^i(p) \\ U_j^i(s_m) \max_{l=1}^{|S|} a_b[s_m, s_l]B_{j+}^i(s_l) & \text{otherwise} \end{cases}$$

Using these equations, the maximum probability that all labels are outputted along the most likely state transition path can be computed as $P^* = \max_l \pi[s_l]\phi_U(s_l, q_1^i)$. Once this is calculated, those states that produce this maximum likelihood can be retrieved using arg max. Then the most likely state transition can be traced by starting with the most likely state for the root, which can be obtained by $q_1^* = \arg\max_l \pi[s_l]\psi_U(s_l, q_1^i)$, and the rest of the states can be found as follows:

$$q_j^* = \begin{cases} \phi_U(s_p^*, q_p) & \text{if} \quad q_j = q_\leftarrow^i(p) \\ \phi_B(s_{j-}^*, q_{j-}) & \text{otherwise} \end{cases}$$

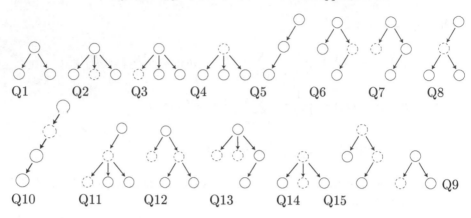

FIGURE 4.14: Fifteen patterns of tree fragments used in OTMM experiments. *Figure reused by permission of the ACM.*

4.4.3.2.1 Experimental results The performance of OTMM was tested in terms of computation time and prediction accuracy using both synthetic and real biological datasets, in comparison with the performance of PSTMM and HTMM, in a supervised learning manner. In particular, for synthetic datasets, up to 15 different fragment patterns were investigated. OTMM was then applied to glycans to find the patterns embedded in them, using a variety of parameter values.

First, the synthetic data set was generated such that positive data sets contained patterns and negative data sets did not. However, the negative data sets needed to have similar background distributions of "family" relationships. Thus, after the positive data sets of trees containing patterns were generated, negative data sets were generated based on the distribution of parent-child labels in the positive data set. Data sets were also ensured to be of consistent size; thus the size of each of the following three datasets (denoted by $|\mathbf{T}|$) were kept the same, and various values of $|\mathbf{T}|$ and $|S|$ were examined. In all experiments, $|\sum| = 10$ and $|V_u| = 20$. Each positive sample contained a pattern as a substructure.

Figure 4.14 shows the 15 tree fragments labeled Q1 to Q15 used in the experiments. The solid circle indicates a fixed label, and the dashed circle indicates a randomly generated label. For example, for pattern Q2, the labels of the parent, the eldest and third siblings are fixed, whereas the label of the second sibling is randomly generated. K different label patterns for each tree fragment were generated in the following manner: First, a random tree was generated by iteratively generating zero to five children and assigning a label to each of the children randomly, until the number of generated nodes reached 20. Second, one of the K label patterns was embedded into the tree. A negative example was also generated using the first step above, except that

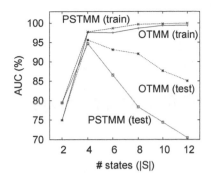

FIGURE 4.15: Performance comparison of OTMM with PSTMM using 100 structures. *Figure reused by permission of the ACM.*

the random generation of labels was based on the distribution of parent-child labels in the positive examples.

The discriminative performance of the models was evaluated using AUC, and the degree of overfitting of OTMM compared to PSTMM was first examined. In this experiment, $|\mathbf{T}| = 100$ and $K = 1$, parameter values where a complex model was expected to easily overfit to a dataset of relatively low complexity. In particular, a model in which all possible state transitions were allowed among all states in S were used. This setting would generate the most complex state transition diagram, called a *fully-connected state model*. Figure 4.15 shows the AUC of OTMM and PSTMM for the training and test datasets, with $|S|$ set to various values in the range of 2 to 12 in order to evaluate increasing model complexity. Note that negative test examples were used for computing the AUC for the training data. For both models, the AUCs for the training data increased with respect to $|S|$ and almost reached 100% when $|S|$ was eight or more. On the other hand, the AUC for the test data decreased with respect to $|S|$ for $|S| \geq 4$. In particular, the AUC of PSTMM went down to just 70% at $|S| = 12$ from 95% at $|S| = 4$. This phenomenon illustrated overfitting to the training data. OTMM had a similar tendency but was less severe. In fact, the AUC of OTMM was always more than 85%, which was approximately 15% better than the worst AUC of PSTMM. Thus, it can be concluded that OTMM reduced the overfitting problem of PSTMM, and consequently OTMM can be considered to be more appropriate for this dataset than PSTMM.

Figure 4.16 illustrates the AUC values of the three probabilistic models for the test examples over a variety of values for $|\mathbf{T}|$ and $|S|$. For all values of $|\mathbf{T}|$ and $|S|$, $K = 3$, fully-connected state models and the tree fragment Q1 was used.

These results illustrate that, for most of the experiments, the performance of OTMM was the best among those of the three models and that the perfor-

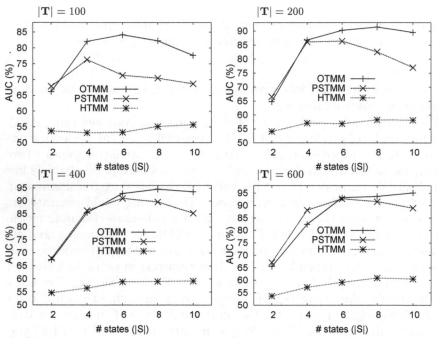

FIGURE 4.16: AUC for fully-connected state models with $K = 3$ and Q1. *Figure reused by permission of the ACM.*

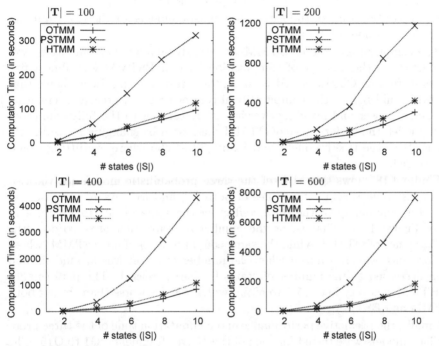

FIGURE 4.17: Computation time for fully-connected state models with $K = 3$ and Q1. *Figure reused by permission of the ACM.*

mance of PSTMM was second, while the performance of HTMM was always much worse than both OTMM and PSTMM. The difference in performance between OTMM and PSTMM was further assessed in more detail. Note that the complexity of the dataset increases with $|\mathbf{T}|$, and that the complexity of the models increase with $|S|$. When $|\mathbf{T}|$ was relatively small (around 100), and $|S|$ was large (around 10), overfitting occurred more easily. However, when $|\mathbf{T}|$ was large (400 to 600), and $|S|$ was smaller (two to six), the AUCs remained at the maximum, indicating that overfitting was avoided. Under optimal conditions, e.g., $|\mathbf{T}| = 400$ and $|S| = 6$, the two models achieved almost the same predictive performance, from which it can be claimed that OTMM has approximately the same predictive power as PSTMM. Figure 4.17 shows the computation time of the three probabilistic models for the test examples over various values of $|\mathbf{T}|$ and $|S|$.

The computation time of OTMM was clearly smaller than the other two. In particular, the amount of computation time of PSTMM was almost five times more than OTMM at $|S| = 10$. In these results, it has been shown that OTMM could reduce the computational cost of PSTMM greatly, keeping its predictive power and avoiding overfitting. Returning to the fully-connected state model, the performance of OTMM was tested by changing K, while still using $Q1$ as the tree fragment and fixing $|\mathbf{T}|$ at 600 where overfitting would be avoided.

Figure 4.18 shows the AUC of the three probabilistic models for the test examples while varying the values of both $|\mathbf{T}|$ and $|S|$. As K increases, so does the complexity of the data, reflecting the same results in Figure 4.18 as in Figure 4.16. That is, as the number of states increases, so does the performance of OTMM, while the same could not be said for PSTMM, whose performance was maximized when the number of states was around six and then decreased as the number of states further increased. The performance of HTMM was very low (55% to 65%), which was far worse than that of both OTMM and PSTMM.

Finally, the predictive performance and computation time of the three probabilistic models were tested for each of the 15 tree fragments, Q1 to Q15. The fully-connected state model and the following parameter settings were used in this experiment: $K = 3$, $|\mathbf{T}| = 600$ and $|S| = 6$, where overfitting did not occur for Q1 using both PSTMM and OTMM.

Table 4.18 lists the AUC of the 15 fragments in this experiment. The best AUC values for each fragment are in bold face. From this table, OTMM performed the best in eight cases, while PSTMM was the best in the remaining seven cases, indicating that the predictive performance of OTMM and PSTMM was comparable, while that of HTMM was much worse.

4.4.3.3 Profile PSTMM

Profile hidden Markov models (Eddy (1998)) were developed as an extension of hidden Markov models (Eddy (1996)) in order to directly extract the

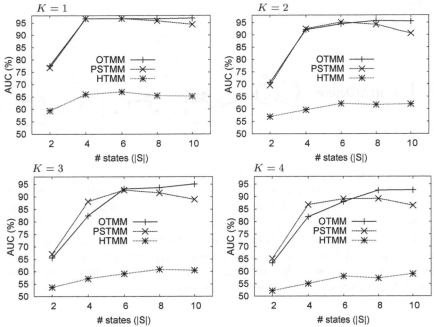

FIGURE 4.18: AUC for fully-connected state models with $|\mathbf{T}| = 600$ and Q1. *Figure reused by permission of the ACM.*

TABLE 4.18: AUC (%) of the three probabilistic models for the 15 tree fragments in Figure 4.14.

Fragment	OTMM	PSTMM	HTMM
Q1	91.2	**93.1**	60.2
Q2	86.3	**90.8**	57.6
Q3	**91.7**	91.3	58.2
Q4	**95.5**	95.2	63.7
Q5	**91.0**	89.9	60.9
Q6	**88.7**	87.8	60.4
Q7	87.1	**88.0**	60.2
Q8	**91.9**	91.1	64.8
Q9	**71.2**	70.2	55.2
Q10	83.3	**86.7**	61.2
Q11	**88.7**	88.3	61.2
Q12	83.0	**85.2**	58.1
Q13	82.6	**83.0**	53.9
Q14	**87.2**	85.6	54.4
Q15	73.9	**75.1**	54.9

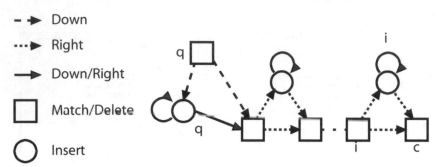

FIGURE 4.19:   A portion of the new profile PSTMM state model with match, insert and delete states. New state transitions are called **Down** for parent-child transitions and **Right** for sibling-sibling transitions. These state transitions are differentiated by different types of dotted lines, and the black lines indicate that both transitions occur between the indicated states. Because match and delete states are always found together, they have also been combined for clarity. Note that this figure is just one parent and its children. These children may also have children, and so on.

profiles trained by the HMM. An introduction to these models are provided in Appendix B. This was achieved by incorporating new types of states whose positions were fixed. A similar improvement could thus be considered for PSTMM/OTMM by adding new types of states, while still maintaining the sibling and parent-child relationships. In order to do this, two types of transitions were integrated in this new model because of the fixed positions of the states. Furthermore, these fixed state positions were also arranged as in OTMM to avoid overfitting problems that could occur when many dependencies exist. This new model was called Profile PSTMM as opposed to Profile OTMM for historical reasons; they were both being developed at around the same time, and the first name remained.

Profile PSTMM extends the previous models to include specialized states such that gaps and substitutions can be accounted for. The extraction of profiles is also straightforward. First, instead of one type of state, there are now three types of states: match, insertion and deletion, denoted as M_i, I_i and X_i, respectively. The positions of the states in the model are also fixed, which helps to reduce computation time and to retrieve the profiles trained more quickly. Like profile HMM, which is described in Appendix B.2.4, profile PSTMM positions match and deletion states at specified positions and insertion states in between. Match states train on the label output probabilities, while insert states output labels with some constant probability and delete states do not output labels at all. Thus the states from the previous models correspond to the Match states in profile PSTMM. Figure 4.19 illustrates the profile PSTMM model. In this figure, new state transitions are illustrated, called *Down* for parent-child transitions and *Right* for sibling-sibling transitions. These state transitions are differentiated by different types of dotted lines, and the black lines indicate that both transitions occur between the indicated states. Because match and delete states are always found together, they have also been combined for clarity. Note that this figure is just one parent and its children. These children may also have children, and so on.

4.4.3.3.1 Probability parameters For simplicity, a value of zero (0) can be assigned to all output probabilities from delete states. Moreover, in order to distinguish between parent-child and sibling relationships, two types of state transitions are also used in profile PSTMM, as in OTMM. Thus profile PSTMM uses the same three probability parameters, π, a and b, as the previous methods, except the model structure is now modified. These probability parameters are again estimated using the same forward, backward, upward and downward probabilities as the previous models, except now taking into consideration the position of the states and the different types of states and state transitions. The forward probability $F_j^i(s_q, s_l)$ is the probability that for node j, all labels of the subtrees of each of the elder siblings are generated, the state of node j is s_l, and the state of parent p is s_q. The following forward probability equations are now defined as follows depending on the state type.

$$F_j^i(s_m, M_l) = \begin{cases} a[\{s_m, -\}, M_l] & \text{if } q_j^i = q_{\leftarrow}^i(p), \\ \sum_{s_k} F_{j-}^i(s_m, s_k)U_{j-}^i(s_k)a[\{s_m, s_k\}, M_l] & \text{otherwise} \end{cases}$$

where state s_k is the state of q_{j-}^i. When s_l is a delete state, M_l above can be replaced by X_l. When s_l is an insert state, the self-loop is taken into consideration as follows.

$$F_j^i(s_m, I_l) = \begin{cases} a[\{s_m, -\}, I_l] & \text{if } q_j^i = q_{\leftarrow}^i(p), \\ \sum_{s_l} F_{j-}^i(s_m, s_l)U_{j-}^i(s_l)a[\{s_m, s_l\}, I_l] & \text{otherwise} \end{cases}$$

where the states at the same position as the insert state s_l are summed.

The backward probability $B_j^i(s_n, s_m)$ is the probability that for node j, all labels of the subtrees of each of the younger siblings and node j are generated, s_m is the state of j, and s_n is the state of its parent. For the backward probability, the same equation can be used for all states s_k, as follows:

$$B_j^i(s_n, s_k) = \begin{cases} U_j^i(s_k) & \text{if } q_j^i = q_{\rightarrow}^i(p), \\ U_j^i(M_k)a[\{s_n, s_k\}, M_l]B_{j+}^i(s_n, M_l)+ & \\ \quad U_j^i(I_k)a[\{s_n, s_k\}, I_k]B_{j+}^i(s_n, I_k)+ & \\ \quad U_j^i(X_k)a[\{s_n, s_k\}, X_l]B_{j+}^l(s_n, X_l)) & \text{otherwise} \end{cases}$$

where s_l is the state of q_{j+}^i.

The upward probability $U_p^i(s_n)$ is the probability that all labels of subtree $t(p)$ are generated and that the state of node p is s_n. The upward probability can be combined for all state types into a single equation. The label output probability when state s_n is a delete state is set to 1.

$$U_p^i(s_n) = \begin{cases} b[s_n, o_p] & \text{if} \quad C_i(q_p^i) = \emptyset \\ b[s_n, o_p] \sum_{s_m} (F_j^i(s_n, s_m) B_j^i(s_n, s_m)) & \text{otherwise} \end{cases}$$

where s_m corresponds to the states of child $x_j \in C(p)$.

Finally, the downward probability $D_j^i(s_l)$ is the probability that all labels of a tree except for those of subtree $t(j)$ are generated and that the state of node q_j^i is s_l. The downward probability parameter is defined as follows.

$$D_j^i(s_l) = \begin{cases} \pi[s_l] & \text{if} \quad j = 1 \\ \sum_n D_p(s_n) b[s_n, o_p] F_j(s_n, s_l) & \text{if} \quad j = q_{\rightarrow}(p) \\ \sum_n D_p(s_n) b[s_n, o_p] F_j(s_n, s_l) \sum_m a[\{s_n, s_l\}, s_m] B_{j+}(s_n, s_m) & \text{otherwise} \end{cases}$$

Here, s_n corresponds to M_n, I_l, and X_n, and s_m corresponds to M_m, I_l and X_m, where s_m is the state of q_{j+}.

4.4.3.3.2 EM algorithm As in previous models, the profile PSTMM probability parameters can be calculated in a backward-breadth-first fashion from leaves to root for upward, forward and backward, and then the downward probability parameter can be calculated from the root back down to the leaves. Thus a similar EM algorithm to calculate the maximum likelihood is used. The pseudocode for parameter estimation is given in Figure 4.20.

Each parameter is calculated not only through the given tree structure but also via the structure of the state model. The pseudocode is simplified and does not specify the details for self-loop transition parameter calculations, but the basic idea is that for insertion states, the state position in the state model does not change. Note that compared to the algorithm for PSTMM, in this new algorithm, not all of the states need to be traversed to call the **find** F, **find** B, **find** U, or **find** D functions since the state to evaluate is given in the arguments. That is, the fixed state positions allow the specification of the states according to position directly. From these changes, it should be apparent that the computation time is drastically decreased, comparable to that of OTMM.

4.4.3.3.3 Likelihood computation The likelihood for a given tree can be computed similarly to PSTMM as the sum of the upward probability multiplied by the initial state probability at the root. The expectation values for π, a, and b are then computed, with which the original values can be updated using the EM algorithm. These expectation values γ, δ and η are calculated similarly to previous models. The computation of $\gamma(\{s_n, s_m\}, s_l)$ is given as an example below.

```
procedure calculate()
  calculate(root,beginState);
  calculateD(root,beginState);
```

procedure calculate(node x, state y)
  /* for all children of x and */
  /* all corresponding state children of y, oldest to youngest */
  for each $c \in C(x)$ **and** $d \in C(y)$ **do**
    calculate(c, d)
  /* from oldest child to youngest child */
  calculateU(eldestNode, eldestState);
  calculateFB(eldestNode, eldestState);

procedure calculateU(node x, state y)
  find $U_x(y)$;
  /* go to immediately younger sibling */
  if x **has younger sibling and** y **has younger state do**
    calculateU(youngerNode, youngerState);

procedure calculateFB(node x, state y)
  find $F_x(parent(y), y)$;
  if x **has younger sibling and** y **has younger state do**
    /* go to immediately younger sibling */
    calculateFB(youngerNode, youngerState);
  else /* go to immediately elder sibling */
    calculateBF(elderNode, elderState);

procedure calculateBF(node x, state y)
  find $B_x(parent(y), y)$;
  /* go to immediately elder sibling */
  if x **has elder sibling and** y **has elder state do**
    calculateBF(elderNode, elderState);

procedure calculateD(node x, state y)
  find $D_x(y)$;
  /* for all children c of x and all children d of y*/
  for each $c \in C(x)$ **and** $d \in C(y)$**do**
    calculateD(c,d)

FIGURE 4.20: Pseudocode for calculating F, B, U and D in profile PSTMM model. *Figure reused by permission of Oxford University Press.*

Defining $H_j(s_n, s_m, s_l) = F_j(s_n, s_m)U_j(s_m)a[\{s_n, s_m\}, s_l]B_k(s_n, s_l)$, for each state type, where q_k is the younger brother of q_j, the calculations are as follows:

$$\gamma(\{s_n, s_m\}, M_l) =$$

$$\frac{\sum_{p:C(p)} D_p(s_n)b[s_n, o_p] \sum_{j \in C(p) \setminus i_\leftarrow(p)} H_j(s_q, s_m, M_l)}{L(\mathbf{T}; \theta)}.$$

Similarly, for the insertion state type:

$$\gamma(\{s_n, s_m\}, I_m) =$$

$$\frac{\sum_{p:C(p)} D_p(s_n)b[s_n, o_p] \sum_{j \in C(p) \setminus j_\leftarrow(p)} H_j(s_n, s_m, I_m)}{L(\mathbf{T}; \theta)}$$

and for deletion:

$$\gamma(\{s_n, s_m\}, D_l) =$$

$$\frac{\sum_{p:C(p)} D_p(s_n) \sum_{j \in C(p) \setminus j_\leftarrow(p)} H_j(s_n, s_m, D_l)}{L(\mathbf{T}; \theta)}.$$

In the maximization step, $\hat{a}$ is updated as as follows:

$$\hat{a}[\{s_n, s_m\}, s_l] = \frac{\sum_u \gamma_u(\{s_n, s_m\}, s_l)}{\sum_u \sum_{l'} \gamma_u(\{s_n, s_m\}, s_{l'})}.$$

Similarly, $\hat{b}$ and $\hat{\pi}$ are updated with their corresponding expectation values. The procedure for computing the expectation values also traverses the state model, so the computation time does not need to iterate through all combinations of states as before.

4.4.3.3.4 Experimental results Profile PSTMM was validated on a synthetically generated data set where profiles are embedded into complex tree structures. This procedure was used to generate three different profile data sets, each of varying complexity, and the ability of profile PSTMM to predict these profiles was assessed. A negative data set was generated similarly to previous validation tests (see Section 4.4.3.2.1). Figure 4.21 illustrates the profiles tested.

For each profile, 50 trees were generated by the following procedure. Take the profile as a tree and randomly generate zero to two levels between the second and third levels, labeling them randomly with symbols from the set

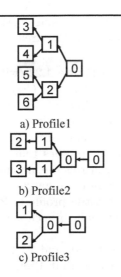

a) Profile1

b) Profile2

c) Profile3

FIGURE 4.21: Synthetic data profiles tested. *Figure reused by permission of Oxford University Press.*

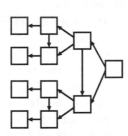

FIGURE 4.22: State model structure for all experiments presented in this work. The Begin state has been omitted. For each node at the first and second levels, $C(i) = 2$. For the third level nodes, $C(i) = 1$, and the leaves are of course $C(i) = 0$. *Figure reused by permission of Oxford University Press.*

TABLE 4.19: Accuracy, precision and AUC values for synthetic data and N-glycan subtype experiments. P1, P2, and P3 represent Profile1, Profile2 and Profile3, respectively. *Table reused by permission of Oxford University Press.*

	P1	P2	P3	High-mannose	Hybrid	Complex
Accuracy	.914	.788	.892	.978	.982	.970
Precision	.843	.974	.926	.882	.904	.882
AUC	.910	.868	.903	.959	.966	.954

$\sum = \{0, 1, \ldots, s\}$ where $s = 7$ for Profile1 and $s = 5$ for Profile2 and Profile3. Additionally, random siblings are added between the leaves up to three children. Taking these 50 trees as the positive data set, 50 additional trees for the negative data set were also generated in order to compare performance. These trees in the negative set were generated based on the parent-child label distributions of the positive set.

The shape of the state model was also fixed as Figure 4.22 (without the Begin state). For each node at the first and second levels, $C(i) = 2$. For the third level nodes, $C(i) = 1$, and the leaves are $C(i) = 0$. This would be sufficient to account for the extra levels in the positive dataset.

As a result, the profiles that were learned from these three data sets are illustrated in Figure 4.23. It is evident from these profiles that the eldest child most strongly learns the data and probably controls the amount of data learned. For example, the profile learned from Profile1 emphasizes 3 and 5 at the eldest leaves of both main branches. Similarly, the profile of Profile2 is learned in the elder main branch, as the younger main branch is essentially random. The same can be said for Profile3.

Finally, the accuracy, precision (at sensitivity of 0.3), and AUC values of these data sets are given in Table 4.19. The reason that Profile2 has the worst performance may be due to the two $1 \leftarrow 0$ linkages that appear in the original profile. This causes the negative dataset to contain this linkage more frequently, thus decreasing the discriminative performance.

4.4.3.4   Structural motifs for lectin recognition

The purpose of this work was to analyze the glycan binding affinity of lectins. In particular, it was preferable to find sialic-acid binding affinity data. However, although sialic-acid binding lectin arrays for glycans have been developed and used for experiments (Bochner et al. (2005); Stevens et al. (2006)) on glycan binding affinity, the glycans spotted on these arrays were mostly trimers, which would not produce very complex profiles. Therefore, the data for glycan binding affinities of galectins that were published in a review by Hirabayashi et al. (2002) were taken. Those galectins that bound to larger and more varied glycans with higher affinity were selected: galectin-3 and galectin-9N. Since the binding affinity data were specified as dissociation constants, the data set was weighted according to binding affinity by proportionately adding more of the glycans that had higher affinity. The binding affinities and corresponding weights of glycans for these two types of galectins are given in Table 4.20. These affinities are the normalized and inverted values from the original disassociation constants so that higher values indicate higher affinity. 30 trees were then randomly selected from the distribution of glycans in this data set. Negative data sets of the same size were also generated based on the parent-child label distribution of the trees in the positive set.

The resulting profiles are given in Figure 4.24. It was not surprising that Gal appeared strongly at the leaves as the nature of galectins is to bind to Gal.

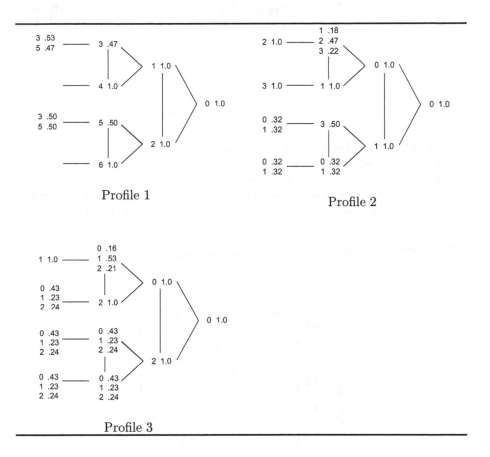

FIGURE 4.23: Profiles learned from synthetic data. Probability values below .20 were omitted. *Figure reused by permission of Oxford University Press.*

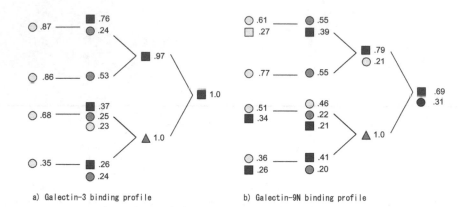

a) Galectin-3 binding profile   b) Galectin-9N binding profile

FIGURE 4.24:   Lectin-binding glycan profiles. Label output probabilities < .20 are omitted. It was not surprising that the galectins appeared strongly at the leaves as the nature of galectins is to bind to galectins. Furthermore, the LacNAc linkage appeared in several of the branches at the leaves, confirming the results in the literature. *Figure reused by permission of Oxford University Press.*

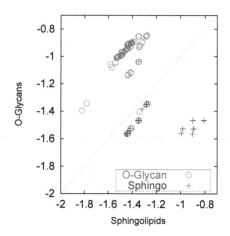

FIGURE 4.25:   Plot of log likelihood values of *O*-glycan and sphingolipid glycans using model trained on *O*-glycans vs. model trained on sphingolipids. *Figure reused by permission of Oxford University Press.*

TABLE 4.20: Binding affinities and weights for Galectin-3 and Galectin-9N. Affinity values are normalized and inverted from the original data by Hirabayashi et al. (2002) such that higher values indicate higher affinity. Abbrev.: NA3: triantennary *N*-Glycan; fuc. NA3: core-fucosylated NA3; NA4: tetraantennary *N*-Glycan; fuc. NA4: core-fucosylated NA4; penta.: pentasaccharide; A-hexa: A-hexasaccharide; LN3: LAcNAc; LN5: (LacNAc)$_5$ *Table reused by permission of Oxford University Press.*

	Gal-3 affinity (weight)	Gal-9N affinity (weight)
NA3	1.28205 (1)	2.6316 (2)
fuc. NA3	1.21951 (1)	2.2222 (2)
NA3 type1	1.08696 (1)	1.6949 (0)
NA4	1.44928 (1)	5.5556 (5)
fuc. NA4	1.40845 (1)	4.3478 (4)
Galili penta.	1.47059 (1)	0.2273 (0)
Forssman penta.	0.16129 (0)	11.111 (11)
A-hexa	1.5873 (1)	3.8462 (3)
LN3	2.85714 (2)	1.2346 (0)
LN5	5.26316 (5)	8.3333 (8)

TABLE 4.21: Performance of lectin-binding glycans for Galectin-3 and Galectin-9N. *Table reused by permission of Oxford University Press.*

	Galectin-3	Galectin-9N
Accuracy	.847	.91
Precision	1.0	.918
AUC	.93	.931

LacNAcs also appeared in several of the branches at the leaves, confirming the results in the literature. The mannose of the Gal-Man linkage is a result of training on the core structure of the N-Glycans in the data set, because of the GlcNAc-GlcNAc linkages at the root. Furthermore, it is noted that the Fuc appearing near the root with 100% probability accounts for the fucosylated core structures of the N-Glycans, and that it usually does not have children. When looking at the state transitions, the transitions out of this state indeed have higher delete transitions compared with the rest of the trained state model (data not shown). Ignoring the descendants of this state, the profiles appeared to capture both the N-Glycan core structures as well as the highly recognized lactosamine structures at the leaves. This result coincides with the results from the original work.

The summary of the accuracy, precision and AUC values for these two models are also presented in Table 4.21, where it is evident that the discrimination of galectin-binding glycans against the negative data set is very high. Thus, it can be claimed that there are indeed patterns that are sibling-dependent in the data which can be directly captured by the profile PSTMM model.

TABLE 4.22:   Tools for the visualization of carbohydrate structures.

Name	Description and accessibility
LiGraph	Given a glycan structure in a textual format, output a schematic figure of the structure.
	http://www.glycosciences.de/tools/LiGraph/
KegDraw	Downloadable application for drawing glycans or chemical compounds in general. Output format in KCF.
	http://www.genome.jp/download/
GlycanBuilder	Both downloadable application and web-based applet available for drawing glycans in CFG, UOXF or as 2D text formats.
	http://www.dkfz-heidelberg.de/spec/EUROCarbDB/
	GlycoWorkbench/builder.html

4.5 Glycomics tools

From the development of glycan structure databases as introduced in Chapter 3, a number of tools for the analysis of glycans were also developed. Those that are freely available are described in more detail in this section.

4.5.1 Visualization tools

Because of the complex nature of carbohydrate structure and biosynthesis, visualization is a crucial element in analysis. Several useful tools have been developed for such analyses, including LiGraph, KegDraw and GlycanBuilder for drawing glycan structures, and GlycoVis for *N*-glycosylation pathway visualization.

One of the first types of tools developed for glycan analysis were visualization tools for displaying glycan structures, especially due to their branched structures, where it is difficult to notate as text. GLYCOSCIENCES.de provided one of the first of these, called LiGraph, which generates graphic files of glycans given text input. KEGG developed the KegDraw application (see Section 3.1.1.2), and GlycanBuilder was developed most recently by a group at Imperial College London (Ceroni et al. (2007)).

4.5.1.1 LiGraph

LiGraph was developed such that images of glycans using a variety of notations could be easily generated from text. A variety of options are also available.

The input screen of LiGraph is given in Figure 4.26. This tool takes as input a carbohydrate structure in text format as a list of names and connections. For example, to specify the Lewis^x structure as in Figure 4.27 (right), the text given to the left should be input, where each alphabet in capital letters represents a variable for the monosaccharide, and each row (except for the first, corresponding to the root), defines a glycosidic linkage. In this example, the L and D conformations are optional, so `a-Fucp` may also be used instead of `a-L-Fucp`. The output of this tool is a new web page displaying the glycan in the specified notation and the corresponding legend.

4.5.1.2 KegDraw

KegDraw is a Java application that can perform queries on the KEGG GLYCAN database through an Internet connection. Figure 4.28 is a snapshot of KegDraw with the *N*-linked glycan core structure drawn. Using the tool panel to the left, a glycan structure can be drawn as a graph, with nodes corresponding to monosaccharides and edges to glycosidic linkages. The buttons with red nodes are pre-defined structures. The button with the letter

```
Enter the names and the connection list

#Enter the template names like this:          5     Pre-fill
#   for b-d-Galp-(1-4)-a-D-Glcp
# A      : a-D-Glcp                          Tutorial
# B > A  : b-D-Galp 1-4
                                             Example

                                              Tips

                                             Reset


Submit   ○ASCII   ○ASCII (numbered)   ○Graph   ⊙Graph      Theme:        Scale:
                                                (numbered)  Heidelberg ▼  1.5
```

FIGURE 4.26: Snapshot of the input screen for the LiGraph tool.

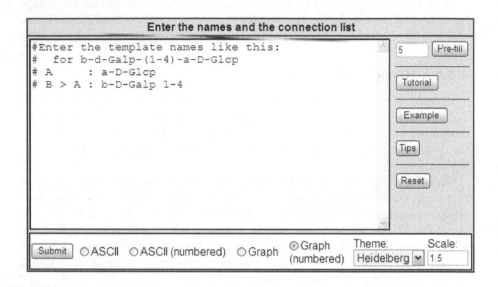

```
A : b-D-GlcpNAc
B > A : a-L-Fucp 1-3
C > A : b-D-Galp 1-4
```

FIGURE 4.27: The Lewis^x glycan motif.

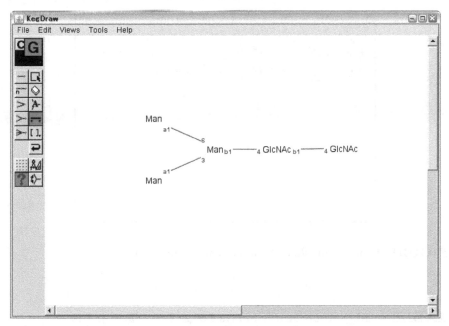

FIGURE 4.28: Example of KegDraw with the *N*-linked glycan core structure drawn.

"A" written on it can be used to draw a residue, and the button below it to draw linkages. User defined residues and linkages may also be specified manually. Finally, the drawn structure can be used as a query to the KEGG GLYCAN database using the **Tools**→Search similar structures menu option. The web interface in Figure 3.2 will be displayed in a new browser window such that the user can specify the search parameters.

4.5.1.3 GlycanBuilder

More recently, the EuroCarb Consortium released GlycanBuilder as an intuitive and rapid tool for drawing glycan structures (Figure 4.29). Glycan-Builder is available both as a stand-alone Java application and as a web-based Java applet. The applet version is used as the glycan structure query interface for GlycomeDB, and as such, it is available to be integrated into any applications requiring an interface for the input of glycans (Ceroni et al. (2007)).

A structure is created in the editor by adding monosaccharides, modifications or reducing-end markers in sequential order. Each addition is performed by selecting the point of attachment and performing the desired action. For example, clicking on the GlcNAc button, a GlcNAc will be automatically displayed along with its mass. With the GlcNAc selected, another GlcNAc can be attached to it by clicking on the GlcNAc button again. The mass of each additional monosaccharide is computed and displayed automatically. Several

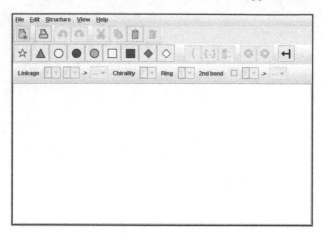

FIGURE 4.29: Snapshot of the GlycanBuilder tool.

TABLE 4.23:   Tools for the analysis of glycan-related pathways.

Name	Description and accessibility
KEGG GLYCAN Composite Structure Map (CSM)	Given one to three monosaccharides for the reducing end, displays all known glycan structures containing those residues in a map, which is linked to the corresponding structures and enzymes. **http://www.genome.jp/kegg-bin/draw_csm**
GlycoVault	An infrastructure for the visualization, analysis and modeling of glycan pathways. **http://glycomics.ccrc.uga.edu**

representations for the structures are available: CFG, UOXF (a format proposed by the Oxford Glycobiology Institute) and 2D text. The structural constituents available cover a comprehensive and continuously updated list of saccharides, substituents, reducing-end markers and saccharide modifications. A library of common structural motifs (core and terminals) is also included such that commonly used structures may be drawn quickly. All the stereo-chemical information about a saccharide, such as anomeric conformation, chirality, ring configuration and linkage position, can also be specified. GlycanBuilder also allows the export of the structure encoded as a GlycoCT string (see Section 2.1.8) or rendered in a standard graphical format (e.g., EPS, PDF, SVG, PNG, JPG).

4.5.2   Pathway analysis tools

4.5.2.1   KEGG Glycan Composite Structure Map (CSM)

The analysis of glyco-related pathways generally consists of the analysis of glycosyltransferases based on the fact that most glycosyltransferases transfer a single residue to an existing residue on the glycan chain. Thus, the biosynthesis of a glycan structure can be surmised from the various glycosidic linkages and residues contained in the structure. In particular, just as "missing" structures could be predicted from two existing structures different by two links, as described in Section 4.3.1, the entire biosynthesis pathway could be similarly analyzed. This was the idea behind the Composite Structure Map (CSM) provided by KEGG. The tool itself is described in Section 3.1.1.

The CSM was built similarly to the glycan variation map (Section 4.2.3), except that the structures A and B were merged only if they differed by exactly one link. Thus those structures that served as substrates for each other were connected such that corresponding glycosyltransferases could be identified. This map was built based on the glycan structures and reaction information in KEGG. Thus each link is mapped to an entry in the KEGG Orthology database, which is a manually curated set of orthologous gene groups found in complete genomes (Kanehisa et al. (2008)).

4.5.2.2   GlycoVault

GlycoVault is an infrastructure for the visualization, analysis and modeling of glycan pathways (Nimmagadda et al. (2008)). It consists of a comprehensive data storage scheme such that data in a number of formats can be stored and retrieved using annotations based on an ontology. Spreadsheets of experimental data are accounted for by GlycoVault, which keeps backups of them in addition to annotations indicating the content and format of the data files. Spreadsheet data may be processed and stored in object-relational databases. Other data in XML or RDF and OWL formats may also be stored in GlycoVault, which provides a common interface to access all data.

GlycoVault also provides two tools for glycomics analysis. GlycoBrowser is a visualization tool for analyzing glycomics data and knowledge over the Web. It can visualize biochemical pathways including reactions and biochemical structures together with relevant experimental data. The second tool is GlyMpse (Glycomics Modeling pathway simulation environment), which is a simulation tool using Hybrid Petri Nets (Matsuno et al. (2003)). Modeling nodes as biochemical entities such as compounds and edges as enzymatic reactions, pathways based on firing delays and enzyme concentrations can be simulated. Simulation datasets consist of metabolic pathways and enzyme kinetics, all stored in GlycoVault.

TABLE 4.24: Tools for PDB data analysis focusing on carbohydrates.

Name	Description and URL
pdb-care	Confirms the validity of carbohydrate residues in a PDB file.
	http://www.glycosciences.de/tools/pdbcare/
pdb2linucs	Given a PDB file, this tool extracts carbohydrate sequences in LINUCS format.
	http://www.glycosciences.de/tools/pdb2linucs/

4.5.3 PDB data analysis

4.5.3.1 pdb2linucs

The GLYCOSCIENCES.de portal (see Section 3.1.2) provides a number of tools for analyzing PDB data, called the Carbohydrate Structure Suite (CSS). One of the fundamental tools comprising CSS is called *pdb2linucs* (Lutteke et al. (2004)), which automatically extracts the carbohydrate structure from a PDB file and generates the structure in LINUCS format (see Section 2.1.5 for details regarding the LINUCS format). This tool utilizes an algorithm that is independent of residue annotation, using only element types and 3D atom coordinates to detect carbohydrate structures, thus overcoming the lack of a standard nomenclature for carbohydrates in PDB and producing consistent carbohydrate structures in LINUCS format. As a result, using this tool, it would be possible to use the extracted LINUCS string to query the GLY-COSCIENCES.de or GlycomeDB databases for structures in PDB. This tool is also used by *pdb-care* to compare the extracted residues with the original residues in the PDB file.

4.5.3.2 pdb-care

Another tool in CSS is *pdb-care* (Lutteke and von der Lieth (2004)). It was found that almost one-third of PDB entries containing carbohydrate entries contained errors, most of which were due to inconsistencies in residue nomenclature or erroneous connection data (Lutteke et al. (2004)). Thus *pdb-care* fulfilled the need for a validity checker in PDB files in regards to carbohydrate structures. This tool compares the detected carbohydrate structures in LINUCS format to the residue assignments as reported in the PDB HET Group Dictionary (http://deposit.pdb.org/het_dictionary.txt). This comparison is performed by using a translation table between these residue descriptions. Three types of residues are considered: monosaccharides, oligosaccharides and residues comprised of both carbohydrates and non-carbohydrates, such as D-Galactose-4-sulphate. Based on this translation table, *pdb-care* reports in detail any problems, inconsistencies or errors detected in a given PDB file.

TABLE 4.25: Tools for the analysis of carbohydrates in 3D space.

Name	Description and URL
GlyVicinity	Computes the statistics of amino acids surrounding a given carbohydrate in 3D space. http://www.glycosciences.de/tools/glyvicinity/
GlySeq	Computes the statistics of amino acids in the neighborhood of glycosylation sites on amino acid sequences. http://www.glycosciences.de/tools/glyseq/
GlyTorsion	Displays all available torsion angles for a glycosidic linkage, ring torsions, omega torsions, etc. http://www.glycosciences.de/tools/glytorsion/
GlyProt	Computes the N-glycosylation of a given PDB structure. http://www.glycosciences.de/modeling/glyprot/
SWEET-II	Constructs 3D models of saccharides from their sequences. http://www.glycosciences.de/modeling/sweet2/

Three types of messages are output by this program: info: describes the type of checks performed, warning: describing non-resolvable discrepancies, and error: describing obviously incorrect assignments. Where possible, when incorrect or ambiguous residues are found, candidates for correct residue names are provided.

Using *pdb-care* and *pdb2linucs*, it then became possible to extract carbohydrate information quickly from PDB data. Three databases were generated using these tools: *GlySeqDB*, *GlyVicinityDB* and *GlyTorsionDB*. *GlySeqDB* contains glycoprotein sequences originating from PDB and SwissProt. This database is used by *GlySeq*, which computes the statistics of amino acids in the neighborhood of *N*- and *O*-glycosylation sites on amino acid sequences. This tool enables the analysis of such information which was previously difficult to obtain.

4.5.4   3D analysis tools

4.5.4.1   GlyVicinity

GlyVicinity is a tool from the CSS for computing the statistics of amino acids surrounding a carbohydrate in 3D space (Lutteke et al. (2005)). It runs as a query tool to *GlyVicinityDB*, which contains lists of distances between amino acids and carbohydrate residues in PDB. *GlyVicinity* then performs statistical analyses of the frequency of amino acids that lie within a user-defined distance up to 10 Å from any carbohydrate residue. Using this tool, one may infer the most important residues for carbohydrate-binding interactions, for example.

TABLE 4.26:   Torsion angles used in the *GlyProt* tool.

Name	Definition	Prioritized frequency of torsion angles
χ_1	$N\text{-}C_\alpha\text{-}C_\beta\text{-}C_\gamma$	180, 200, 300, 280, 60, 80, 40, 220, 320
χ_2	$C_\alpha\text{-}C_\beta\text{-}C_\gamma\text{-}0$	340, 320, 20, 0, 40, 60, 280, 80, 280
Ψ_n	$C_1\text{-}N_1\text{-}C_\gamma\text{-}C_\beta$	160, 180, 200
φ_n	$O_5\text{-}C_1\text{-}N_1\text{-}C_\gamma$	260, 280, 240, 220, 300

4.5.4.2   GlyTorsion

GlyTorsion is a tool from the CSS for querying *GlyTorsionDB*, which is a database of torsion angle values of all carbohydrate linkages in PDB. It also contains ring torsion angles of single monosaccharides, omega torsion angles for exocyclic hydromethyl-groups, side chain torsion angles of asparagine residues involved in *N*-linked glycosylation and the torsion angles of *N*-acetyl groups attached to carbohydrate rings. *GlyTorsion* produces histograms of the distribution of angles for specific structural features. Thus, the most preferred angles can be evaluated for a particular residue, for example.

4.5.4.3   GlyProt

GlyProt takes as input a PDB structure and computes the potential *N*-glycosylation sites on the protein based on spatial accessibility. Because not all asparagine residues are *N*-glycosylated, this tool assists in predicting those that are most likely to be glycosylated. It can then generate 3D models of glycoproteins with glycans specified by the user, allowing one to analyze how the physicochemical parameters are affected by varying protein glycoforms (Bohne-Lang and von der Lieth (2005)). This tool first highlights the potential *N*-glycosylation sites based on the motif Asn-X-Ser/Thr, where X is any amino acid except for Pro. The internal coordinates of any existing glycans in the PDB file are displayed, where internal coordinates refer to the distance between the *N* of the Asn-sidechain and the C1 of the attached β-D-GlcNAc and the torsion angles determining the orientation of the glycan.

To compute the spatial accessibility, four types of torsion angles are used to define the orientation of the glycan relative to the protein: χ_1, χ_2, Ψ_n and φ_n. These torsion angles are defined in Table 4.26 and can be obtained easily using the *GlyTorsion* tool. First, the tri-mannose core is connected to the protein and all possible angle sets are tested. The frequency of occurrence of the four types of torsion angles as listed in Table 4.26 is used to orient the *N*-glycan core structure. Next, the program determines if the glycan structure overlaps the protein, and if so, the next orientation is tested. This procedure is repeated until a structure with no or little overlap has been found. If no such structure can be found, it is assumed that the particular glycosylation site is spatially inaccessible.

The input to GlyProt may be a PDB ID or a text file in PDB format. After submission of the input, the following computations are performed: (1) se-

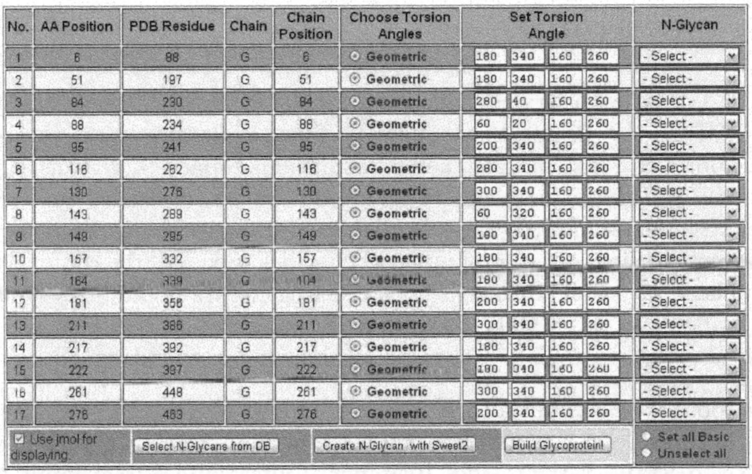

No.	AA Position	PDB Residue	Chain	Chain Position	Choose Torsion Angles	Set Torsion Angle				N-Glycan
1	6	88	G	6	○ Geometric	180	340	160	260	- Select - ⌄
2	51	197	G	51	⊙ Geometric	180	340	160	260	- Select - ⌄
3	84	230	G	84	○ Geometric	280	40	160	260	- Select - ⌄
4	88	234	G	88	⊙ Geometric	60	20	160	260	- Select - ⌄
5	95	241	G	95	○ Geometric	200	340	160	260	- Select - ⌄
6	116	262	G	116	⊙ Geometric	280	340	160	260	- Select - ⌄
7	130	276	G	130	○ Geometric	300	340	160	260	- Select - ⌄
8	143	289	G	143	⊙ Geometric	60	320	160	260	- Select - ⌄
9	149	295	G	149	○ Geometric	190	340	160	260	- Select - ⌄
10	157	332	G	157	⊙ Geometric	180	340	160	260	- Select - ⌄
11	164	339	G	164	○ Geometric	180	340	160	260	- Select - ⌄
12	181	358	G	181	⊙ Geometric	200	340	160	260	- Select - ⌄
13	211	388	G	211	○ Geometric	300	340	160	260	- Select - ⌄
14	217	392	G	217	⊙ Geometric	180	340	160	260	- Select - ⌄
15	222	397	G	222	○ Geometric	190	340	160	260	- Select - ⌄
16	261	448	G	261	⊙ Geometric	300	340	160	260	- Select - ⌄
17	276	463	G	276	○ Geometric	200	340	160	260	- Select - ⌄

☑ Use jmol for displaying.　　Select N-Glycans from DB　　Create N-Glycan with Sweet2　　Build Glycoprotein!　　○ Set all Basic　○ Unselect all

FIGURE 4.30: Using GlyProt, the input screen for specifying the *N*-glycans to attach to the inputted PDB structure.

quence extraction and computation of potential N-glycosylation sites, (2) survey of any MODRES entries in the PDB file, indicating pre-specified protein modifications, (3) search for any carbohydrates defined in the PDB file, and (4) computation of the spatial accessibility of the potential N-glycosylation sites. All this information is first presented upon submission of the input. Next, the user is prompted to specify N-glycans to attach to the protein structure, as in Figure 4.30. The options for the N-glycans include Basic, Oligomannose, Hybrid, Complex, Poly-N-acety., and Very Large. In most cases, Basic may suffice, unless it is known that the structure contains very large or complex glycan structures. Upon clicking the "Build Glycoprotein!" button, a Jmol viewer will be displayed showing the glycosylated protein structure.

4.5.4.4   SWEET-II

SWEET-II is a web service that automatically generates carbohydrate structures in 3D space (Bohne et al. (1999)). This application uses a systematic search approach for exploring the conformational space of glycans via rotations about the glycosidic bonds. This in addition to rule-based approaches is applied to generate 3D structures of glycans. First, the given glycan structure is constructed using a library of monosaccharides. Second, the conformational space of each glycosidic linkage is explored to generate preliminary conformations (Imberty et al. (1990, 1991); von der Lieth et al. (1997)), which are optimized using a complete molecular mechanics force field.

This tool takes as input a carbohydrate structure which can be specified in a form or directly in CarbBank format (see Section 2.1.3). In particular, the interface to SWEET-II provides Beginner, Expert and Direct input modes, where a simple form is provided for Beginner mode, a more complex form is provided for Expert mode, and a single text area is provided for Direct input mode such that CarbBank formatted data can be inputted. In the resulting page, the LINUCS code for the structure is displayed along with a 3D image of the input. Information regarding the Ψ and φ conformational maps used to find the best conformation as well as errors are provided as a text file. The resulting oligosaccharide can be visualized using a molecular display program such as Rasmol (Sayle and Milner-White (1994)) or Webmol (Walther (1997)).

4.5.5   Molecular dynamics

The flexibility of glycan structures oftentimes necessitates molecular dynamics simulations in order to truly understand their functions. Thus some of the simulation software that have been developed to facilitate such analyses are described here.

TABLE 4.27: Tools for molecular dynamics simulations.

Name	Description and URL
Dynamic molecules	Web-based tool for performing molecular dynamics simulations of not only oligosaccharides but other molecules such as proteins and DNA as well.
	http://www.md-simulations.de/manager/
GlycoMapsDB	By inputting a disaccharide structure, this tool automatically generates conformational maps from long-term molecular dynamics simulations using Dynamic molecules.
	http://www.glycosciences.de/modeling/glycomapsdb/

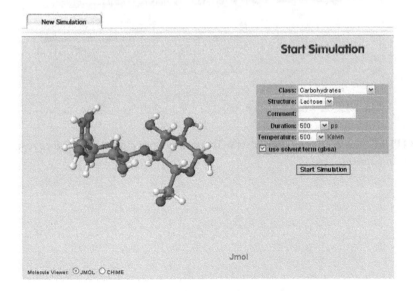

FIGURE 4.31: Snapshot of the Dynamic molecules tool in Beginner mode.

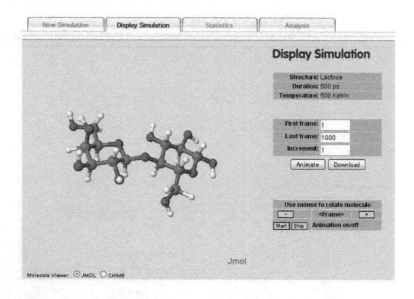

FIGURE 4.32:   Snapshot of the Dynamic molecules simulation results in Beginner mode.

4.5.5.1 Dynamic molecules

Dynamic molecules is the first web site to offer molecular dynamics simulations to the public (Frank et al. (2003)). It allows any user with little to no experience to run simulations over the web. This tool allows users to use beginner mode or expert mode, the latter providing more options for fine-tuning the simulation. As an example, Figure 4.31 is a snapshot of the Dynamic molecules tool in Beginner mode. Here, the lactose structure is selected, and options for the time frame over which to run the simulation, the temperature, and the option to include a solvent or not is provided.[6] The resulting simulation is thus displayed as in Figure 4.32.

In contrast, the Expert mode provides options to specify user-defined carbohydrate structures, such as in Figure 4.33, in addition to the execution of commonly used scripts to perform dynamics simulations that expert users would normally use.

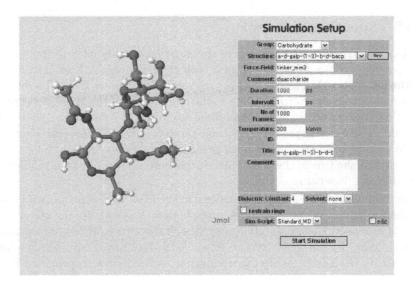

FIGURE 4.33: Snapshot of the Dynamic molecules simulation results in Expert mode.

---

[6]Note that at the time of this writing, this tool is fully functional under Mozilla or Netscape, and that the Java or Chime plug-ins should be installed.

4.5.5.2   GlycoMapsDB

Using the Dynamic molecules tool, it was then possible to perform simulations of small carbohydrates. Thus, a database of the results of such simulations was created in the form of GlycoMapsDB (Frank et al. (2007)). This database contains conformational maps of more than 2500 di-, tri-, up to pentasaccharide fragments contained in *N*- and *O*-linked glycan structures from CarbBank (Doubet et al. (1989)). The 3D structures of these fragments were generated using SWEET-II (see Section 4.5.4.4), and a molecular mechanics force field (MM3) was applied to calculate the trajectories at 1000 Kelvin. Simulations of 10ns for disaccharides and 30ns for larger structures were performed and stored in the database.

GlycoMapsDB can thus be queried by specifying a glycan structure in extended IUPAC format (see Section 2.1.3). The matching structures are listed with previews of their conformational maps. It is also possible to compare differences between the conformational maps of two structures.

4.5.6   Spectroscopic tools

Mass spectrometry (MS) is one of the most commonly used technologies to characterize glycan structures from a biological sample. A brief introduction on the concept of MS techniques is provided in Appendix C.1. The difficulty in using MS to characterize glycans is the complexity of the structures and the resulting spectrometry data. Thus some useful tools have been developed to ease this process (GlycoFragment, and GlycoSearchMS). Several algorithms for MS annotation of glycans have been introduced in Section 4.3.4. These are distinguished from the tools introduced here due to public availability at the time of this writing.

Additionally, nuclear magnetic resonance (NMR) experiments can be used to identify complex carbohydrates unambiguously, down to the stereochemistry of monosaccharides and linkage types. Thus two methods for the identification of carbohydrate structures from NMR data have also been developed (CASPER, GlyNest). A brief introduction to NMR is also provided in Appendix C.2.

4.5.6.1   GlycoMod

GlycoMod is provided by the Expasy server and is one of the first tools to assist in the annotation of MS data of glycoproteins or free oligosaccharides (Cooper et al. (2001)). It predicts the possible glycans attached to proteins from their experimentally determined masses; it compares the mass of a potential glycan against a list of pre-computed masses of glycan compositions. Furthermore, since it is connected to Expasy, the experimentally determined masses will be matched against all the peptides in the SWISS-PROT or TrEMBL databases that have the potential to be *N*- or *O*-glycosylated.

TABLE 4.28: Spectroscopic tools

Name	Description and URL
GlycoMod	Predicts possible oligosaccharide structures occurring on proteins based on their experimentally determined masses. This is performed by comparing the mass of a potential glycan to a list of pre-computed masses of glycan compositions. http://www.expasy.ch/tools/glycomod/
GlycoFragment	Generates all theoretically possible MS relevant fragments of oligosaccharides such that they can be compared against existing MS data. http://www.glycosciences.de/tools/GlycoFragments/
GlycoSearchMS	Takes as query a mass spectrum and returns the most likely spectra from a database of theoretically calculated spectra. http://www.glycosciences.de/sweetdb/ start.php?action=form_ms_search
GlycoWorkBench	Java application to assist in the annotation of mass spectra of glycans. http://www.dkfz-heidelberg.de/spec/ EUROCarbDB/GlycoWorkbench/
CASPER	Increment-rule approach to estimate ^{1}H- or ^{13}C spectra of a given glycan structure. http://www.casper.organ.su.se/casper/
GlyNest	Spherical environment encoding approach to estimate ^{1}H- or ^{13}C spectra of a given glycan structure. http://www.glycosciences.de/sweetdb/ start.php?action=form_shift_estimation

4.5.6.2 GlycoFragment

GlycoFragment takes as input a carbohydrate structure and produces a list of all possible fragments and ion adducts (Lohmann and von der Lieth (2003)). The input structure should be specified in extended IUPAC format (see Section 2.1.3), with some additional options to specify residues as hex for hexose structures, for example, when the specific monosaccharide is unknown. The input may also include derivatives formed by reductive amination or persubstituted derivatives of carbohydrates.

The output is a list of B- and Y-type fragments ordered by increasing mass, thus providing a general overview of the possible fragments. The less frequently occurring C- and Z-fragments as well as all possible rings fragmentations can also be listed. The results also contain the option to "View as structure," where the structure is displayed as in extended IUPAC format, and moving the cursor over the structure displays the associated fragments.

4.5.6.3   GlycoSearchMS

GlycoSearchMS takes as input a mass spectra and returns the most closely matching spectra from a database of theoretical spectra (Lohmann and von der Lieth (2004)). GlycoSearchMS computes the masses of all A-, B-, C-, X-, Y- and Z-fragments and assigns peaks according to Domon and Costello (1988). The tool then continues to compare each peak of the input with the computed fragments of all structures in the database. The following equation is used to score the peaks in the database:

$$MS_{score} = \frac{\sum_1^n \{1 - (|P_s - P_r|/Err)\}}{n_{input}} \times 100$$

where n is the number of input peaks, P_s is the mass-to-charge ratio (m/z) input peak, P_r is the m/z reference peak from the database, and Err is the tolerance for the number of matched peaks in mDa units.

4.5.6.4   GlycoWorkBench

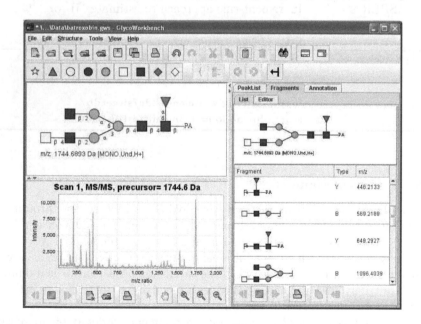

FIGURE 4.34:   Snapshot of the GlycoWorkBench tool.

GlycoWorkBench not only consists of a drawing tool for drawing glycan structures (called GlycanBuilder, introduced in Section 4.5.1.3), but tools for annotating mass spectrometry data with glycan structures (Ceroni et al. (2008)). Figure 4.34 is a snapshot of this tool (taken from http://www.dkfz-heidelberg.de/spec/EUROCarbDB/GlycoWorkbench/images/screenshot.jpg). The top left panel is the GlycanBuilder panel, where glycan structures can be specified using the appropriate buttons in the panel along the top. The mass of the drawn structure is automatically computed according to the type of per-substitution (such as per-methylation, per-acetylation, etc.), if any, the identities and quantities of ion adducts (such as H^+, Na^+, K^+ or Li^+) and the neutral exchanges. The resulting m/z value is displayed below the structure.

A peak list can be specified in the panel on the right. Data can be loaded from a tab-delimited text file, or mass and intensity values can be entered manually. Raw spectra can be loaded from a file in any of the standard XML or even vendor-specific formats. The data is displayed in the Spectra panel on the bottom left, which can be panned or zoomed for more detail. Users can then select m/z values directly from the spectrum and add them to the peak-list.

Given a glycan structure, GlycoWorkBench can generate all topologically possible fragmentations of the precursor molecular ion, applying both multiple glycosidic cleavages and cross-ring fragmentations. The resulting list of fragments can be viewed in the Fragments list panel, which contains for each row the fragment structure, the fragment type, the mass-to-charge ratio given the ion adducts, the identities and quantities of ion adducts, the neutral exchanges if any, and the mass of the fragment without any adducts. For automatic annotation, the list of fragments generated above is compared with the peak list to find the most closely matching pairs. Several utilities are available in GlycoWorkBench to facilitate the annotation of spectra.

4.5.7 NMR tools

Nuclear magnetic resonance (NMR) provides a means to identify glycan structures from a sample that is complementary to MS techniques. There are two tools that are publicly available which allow users to analyze NMR data of carbohydrates: CASPER and GlyNest.

4.5.7.1 CASPER

CASPER is a program developed and hosted by Stockholm University (Loss et al. (2006)). This program uses an increment rule based approach using the chemical shifts of free reducing-end monosaccharides which are altered according to the attached residues in a glycan structure. Glycosylation shifts for a linkage can be obtained from the chemical shifts of a disaccharide by subtracting the chemical shifts of the component monosaccharides. Correspondingly, corrections can be obtained by subtracting the monosaccharide and glycosylation shifts from the chemical shifts of a trisaccharide.

FIGURE 4.35:   Structure used as an example to describe CASPER.

TABLE 4.29:   Computation procedure of CASPER for the Glc residue of the glycan in Figure 4.35.

Step #		C1	C2	C3	C4	C5	C6
1.	Glc	96.84	75.20	76.76	70.71	76.76	61.84
2.	Fucα1-3Glc	-0.21	0.31	7.19	-1.38	-0.02	0.01
	Galβ1-4Glc	-0.17	-0.22	-1.46	9.19	-1.14	-0.56
	Result	96.46	75.29	82.49	78.52	75.60	61.90
3.	Corrections	0.27	1.21	-4.20	-4.60	0.73	-0.37
	Final result	96.73	76.50	78.29	73.92	76.33	60.92

The procedure first evaluates the chemical shifts of the individual monosaccharides of a glycan. The glycosylation shifts for each linkage are then added, and then any corrections for vicinal substitution are added in the last step. All these computations are performed for each carbon atom of a monosaccharide. For example, given the structure in Figure 4.35, these three steps would be computed for the glucose residue as in Table 4.29.

4.5.7.2 GlyNest

GlyNest was developed and is hosted by the German Cancer Research Centre as part of the GLYCOSCIENCES.de portal. Using a spherical environment encoding scheme, the chemical shifts of glycans are estimated (Loss et al. (2006)). This encoding scheme focuses on each atom of a glycan structure and applies the following rules to take account of the fact that closely located atoms would affect the chemical shift of a given atom more than atoms farther away.

For example, given the N-glycan core structure, one can estimate the chemical shifts of the first mannose residue (linked to GlcNAc) using a spherical environment encoding scheme as illustrated in Figure 4.36. This figure demonstrates the encoding scheme for each C-atom of this mannose residue. The following rules were applied to order the list of attached residues: (1) the connected residues are ordered according to increasing distance (in terms of number of bonds) from the atom to be encoded and (2) if two distances are equal, the residue attached to the C atom with the smaller ring-atom number has higher priority.

```
C1:(1-4)B-D-GLCPNAC:(3+1)A-D-MANP:(6+1)A-D-MANP
C2:(1-4)B-D-GLCPNAC:(3+1)A-D-MANP:(6+1)A-D-MANP
C3:(3+1)A-D-MANP:(1-4)B-D-GLCPNAC:(6+1)A-D-MANP
C4:(3+1)A-D-MANP:(6+1)A-D-MANP:(1-4)B-D-GLCPNAC
C5:(6+1)A-D-MANP:(1-4)B-D-GLCPNAC:(3+1)A-D-MANP
C6:(6+1)A-D-MANP:(1-4)B-D-GLCPNAC:(3+1)A-D-MANP
```

FIGURE 4.36: Spherical environment encoding scheme, as used by GlyNest, for the central mannose residue in the N-glycan core structure.

Taking all the structures in the GLYCOSCIENCES.de database, the chemical shifts for the available atoms were generated and stored together in a shift-environment table. This table is used to estimate the chemical shifts of a given glycan by generating the corresponding codes for each atom of the input molecule and looking it up in the table. Because it may be possible to retrieve more than one hit with differing shift values, the result includes the mean, standard deviation, minimum and maximum values of the retrieved hits. If no match is found for a complete environment code, then residues are consecutively removed from the end of the code, and the search is repeated until a hit is found.

In a further step, GlyNest uses the results from CASPER to integrate them into the returned results. Given an input molecule, GlyNest first converts the structure to CASPER line notation and sends this input to CASPER over the Internet. CASPER then computes the NMR spectrum, labels the residues and atoms, and returns an XML encoding of the assignments, which is sent back. Finally, GlyNest parses the XML and integrates the CASPER shifts into its output list.

Chapter 5

Potential Research Projects

This chapter will introduce the research topics that are currently needed in this field of glycomics/glycome informatics. These potential projects are only described briefly as they are just suggestions of where one may start in their research. As is in most cases, this field is continually evolving, and it is recommended that readers looking for projects in glycobiology assess the literature after reading this chapter.

5.1 Sequence and structural analyses

The field of glycobiology not only focuses on glycan structures, but much research also focuses on the glyco-genes that are involved, such as glyco-enzymes and lectins. As mentioned previously in Section 4.3.2, however, the carbohydrate binding motif is very short (DxD), and several groups have attempted to classify these genes by patterns in sequence. The CAZy database is one outcome of such analyses, based on 2D and 3D structures.

This area of research is still open-ended, however. Ideas for different methodologies for characterizing these genes, perhaps not only in the field of glycobiology, are highly welcomed. Kernel methods and other data mining techniques may also be considered.

The basic tree structure alignment algorithm was introduced in Section 4.2.1, upon which a glycan score matrix was developed (Section 4.2.2). The drawback to these algorithms is that they are based on the existing data. However, it cannot be assumed that all carbohydrate structures in the glycome have been discovered and registered into these databases. Thus algorithms for the prediction of all possible glycan structures would be an adventurous project to start with. Data mining methods for capturing the patterns by which glycan structures are synthesized in a particular organism or cell type may be used for such predictions.

The structure of glycan polymers and their functions are as of yet untouched by the bioinformatics community. It is known that the integrity of the peptidoglycan surrounding bacterial cells are crucial for their survival. It has also been shown that polymers containing 6'-sulfated sialyl Lewis^x structures bind with Siglec-8 on human eosinophils in a selective manner (Hudson et al. (2009)). These structures may provide clues to the multivalency issue as well.

The evolution of glycan diversity is another area of research interest. Varki et al. (2008) provide a comprehensive introduction to the current knowledge of this area, which is in fact very little. Similar to the Composite Structure Map or variation map developed by KEGG GLYCAN, the analysis of glycan diversity in terms of evolution is a possible research project that would be quite useful to the glycobiology community.

5.1.1   Glycan score matrix

The glycan score matrix algorithm may be improved by incorporating other sources of information. For example, the biosynthesis pathways of glycan structures is another source that may be incorporated in order to further generate better score matrices. Instead of counting the frequency of glycan linkage alignments directly, the enzymes used to generate the corresponding linkages may be counted.

This will entail the filtering of the data set such that only those structures found in the species which contains the corresponding enzymes are considered. That is, the glycan structures being analyzed in generating a score matrix in this manner will need to be those that can be found in the species for which the enzymes are being considered. This organism information for each data structure is readily available in the GlycomeDB database (Section 3.1.7).

5.1.2   Visualization

Visualization tools for the comprehensive analysis of the biosynthetic pathways of *N*-glycans were introduced in Sections 4.3.3 and 4.5.2. *N*-glycans are the most well-defined of the glycan classes and thus many tools have been developed for these structures. However, there are many other classes of glycans including *O*-glycans, glycosphingolipids and glycosaminoglycans that are also known to be important for biological function. Some databases and numerous publications provide information regarding these structures. It is a matter of (1) organizing this data into a database and (2) developing models such that these classes of glycans may also be analyzed at near, if not at, the same level as *N*-glycans.

The conformation of glycan structures is also an area of active research, with molecular dynamics (MD) simulations often using much computing power to analyze glycan-protein interactions (see Section 4.5.5). The flexibility of glycan structures complicates the analysis of these interactions. Thus the environment in which these interactions take place often affects binding affinity. Such data has yet to be accumulated, albeit several groups have been conducting such simulations for years. This goes hand-in-hand with the MD analysis of lectins and glyco-enzymes. 3D analysis tools (Section 4.5.4) and databases

(Section 3.1.5) may be possible starting points for developing algorithms and tools for such research.

5.2 Databases and techniques to integrate heterogeneous data sets

The major focus of the last few years has been the development of a consistent glycan structure database. These databases have also been integrated into a single interface via GlycomeDB. However, this information is only part of the big picture. The functions of glycans are found mainly in their interactions with other biomolecules. Thus glycan binding affinity data and lectin arrays have been popular tools for glycomics analysis.

There are many small-scale databases containing useful information related to glycans, as introduced in Section 3.5. The integration of all these databases into a single portal would be a close to impossible task. A distributed infrastructure may be a more feasible project to accomplish the same goal of integrating heterogeneous data. In fact, the Japanese government has encouraged the development of the Integrated Database Project, which is being organized by the Database Center for Life Science (DBCLS) in Japan (http://lifesciencedb.mext.go.jp/en/). The main goal of this project is to support the integration of all life science databases in Japan in a distributed manner. Thus, individual databases maintain their web interfaces, but their data can be accessed via an integrated web interface which links to other related information in other databases. As a part of this project, the National Institute of Advanced Industrial Science and Technology (AIST) has taken the lead in developing an integrated interface for glycan-related databases in Japan. This project is still in the works, and the goal for completion is aimed towards 2010.

Another related project was the EuroCarbDB project, whose goal was the integration of carbohydrate-related databases in an integrated manner as well. This project was focused on mass spectrometers in particular due to the large amounts of data generated by individual groups based on this technology. The bottleneck came by the lack of access to this data. GlycoWorkBench was one part of the EuroCarbDB project, and GlycomeDB will become an integrated part of the central database.

In addition to these projects in Japan and Europe, the Consortium for Functional Glycomics (CFG) may also be considered a form of a portal for carbohydrate-related data. The continuous addition of new glycan affinity and profile data is a valuable source of integrated information.

The integration of all these databases will require agreement by all parties on a standard for data exchange, for which GLYDE-II has been proposed

(Packer et al. (2008)). This is a start, but further integration such as through the development of web service workflows are still in the early stages, and more discussions are needed by the parties involved in setting guidelines for such development.

5.3   Automated characterization of glycans from MS data

Several methods for the automatic characterization of glycan structures from MS spectra were introduced in Section 4.3.4. These methods are currently not implemented as tools in the public domain. Furthermore, standardized tests are necessary for benchmarking these methods against all of the varieties of mass spectrometers that may generate spectra for glycans and glycoconjugates. This is an area of high demand not only for glycobiologists, but also for database managers who wish to build upon their current datasets.

There are also a wide variety of MS technologies, only partly covered in Appendix C. Algorithms and tools that can handle these other types of data generated by different MS technologies are needed.

5.4   Prediction of glycans from data other than MS

A method for the prediction of glycan structures from microarray data was introduced in Section 4.3.1. This method was developed in the hopes that a faster method of glycan profiling may be established compared to MS annotation. Similar methods taking into consideration glyco-enzyme expression and substrate specificity are possible. Known information regarding existing biosynthesis pathways may also be sources of information. Tools such as GlycoVis (Section 4.3.3.1) may be utilized in such analyses. Data mining methods that can take advantage of glycan profiling data from the CFG, for example, may be used for glycan structure prediction as well.

5.5 Biomarker prediction

The prediction of glycan biomarkers is also an area of interest by the glycobiology community. As was described as the motivation for developing Profile PSTMM (Section 4.4.3), the mechanism of recognition of glycan structures is still not clearly understood. There are several glycan structures that are known markers, and many of the same structural motifs appear as biomarkers for different diseases. The complexity of recognition is something that may be difficult to capture with structural data alone. Thus the use of kernels to combine multiple data types for classification is another area of high interest. The issue lies in the determination of the appropriate types of data, especially considering the cell- and tissue-specificity of glycans. Of course this will involve close communication with experimentalists. It is not assumed that a single algorithm will be able to handle all such classifications, and that sample-specific methods will probably be more robust and useful to the glycobiology community.

Furthermore, multivalency was described in Section 2.2 to explain how lectins bind weakly with single glycans, but strongly with multiple glycans on the cell surface. This issue has not been addressed from a bioinformatics viewpoint. However, the recent development of lectin arrays have accumulated data that may aid in the development of predictive algorithms. This area of research may be tied to biomarker prediction as well.

5.6 Systems analyses

Sequence and structural analyses may also be combined into a systems-based research project, whereby the glycome of a specific species is the focus, for example. The systems-based visualization tools described in Section 4.3.3.1 may be a starting point for such a project. The CFG has also provided many tissue-specific and cell-specific data that may be utilized.

The relationships between virus-host and pathogen-host is also an interesting area of research in terms of glycobiology. It is known that viruses acquire glycosylation from the host. Although there are some exceptions to the rule, with some viruses containing genes encoding unusual glycosyltransferases, some phage viral glycosyltransferases can modify their surface antigens to change the serotype of their host bacteria or to even glycosylate their own DNA to block it from degradation by restriction enzymes. It is also known that pathogens mimic the glycosylation patterns of the host to aid in better survival through manipulating host immunity. Thus many systems-level analyses (and beyond) are possible.

5.7 Drug discovery

Section 1.4 described the various congenital disorders of glycosylation, or CDGs, that are known, caused by mutations in glycosyltransferases necessary for the biosynthesis of glycans. Thus treatments such as by the oral treatment of monosaccharide supplements have been attempted. However, most CDGs are untreatable. There is hope, however, from a surprising discovery in the muscular dystrophies, whereby the overexpression of a particular glycosyltransferase prevented the development of a pathology caused by defective glycosylation of α-dystroglycan or the loss of dystrophin itself. Thus, it was shown that treatments via the modification of glycosylation patterns may be a possibility (Freeze (2006)).

The discovery of drugs that can bind to particular glycan patterns, thus preventing the promotion of diseases, is also an important area of research. The GlycoEpitope DB (Section 3.5.1) is one possible source of invaluable information regarding carbohydrate antigens and antibodies.

Appendix A

Sequence Analysis Methods

This appendix chapter will describe the basic bioinformatics methods that serve as the foundation for this book. The dynamic programming method for sequence alignment and BLOSUM score matrix algorithm are introduced.

A.1 Pairwise sequence alignment (dynamic programming)

A.1.1 Dynamic programming

In order to introduce the dynamic programming algorithm, a basic algorithmic problem will first be described, and dynamic programming will be used to solve it. This problem is called the Manhattan Grid Problem, and starts with a grid representing a map with various sightseeing locations, as in Figure A.1. Here, the problem is to start from the top-left point (START) and travel to the bottom-right point (END) while visiting the maximum possible number of sightseeing locations while only moving right or down (no backtracking allowed).

This problem can be formulated by first adding weights to each edge corresponding to the number of sightseeing locations along the edge. Thus the problem is to maximize the sum of the visited edge weights along a path from the START position to the END position, as illustrated in Figure A.2.

The grid in this problem can be numbered along the x- and y-axes, with the START position being $(0,0)$ and the END position being (n,m) depending on the number of rows and columns. We can then solve this problem in a recursive fashion by determining the method to travel to a particular position (i,j) based on the possible positions from which that position can be visited plus the sum of the weights up to that previous position. An example is given in Figure A.3, where it is assumed that the maximum sum of edge weights of the path from $(0,0)$ to $(i-1,j)$ is 10 and that of the path from $(0,0)$ to $(i,j-1)$ is 12 and the weight of the edge from $(i-1,j)$ to (i,j) is 4 while the weight of the edge from $(i,j-1)$ to (i,j) is 1. In this case, the maximum sum of edge weights from $(0,0)$ to (i,j) would be 14 where the path from $(0,0)$ to $(i-1,j)$ and the edge from $(i-1,j)$ to (i,j) is taken. If this procedure is taken from the START position $(0,0)$ where all previous maximum paths are

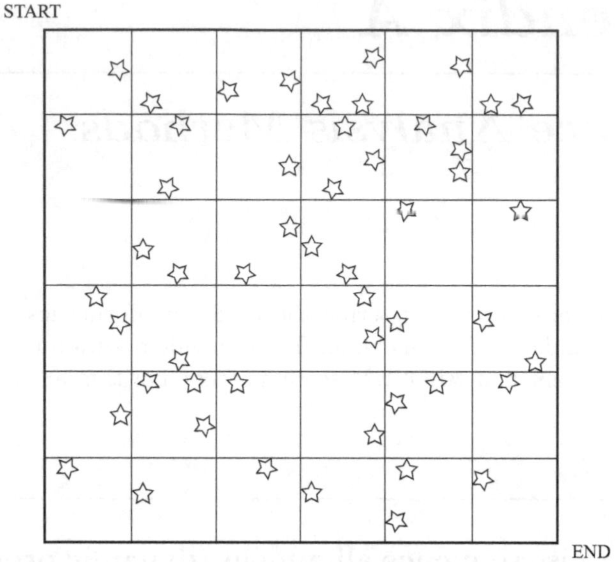

FIGURE A.1: The starting point of the Manhattan Grid Problem, where a grid is given, representing a map of sightseeing locations to potentially visit, and the problem is to travel from the top left corner (START) to the bottom right (END) while visiting the largest number of sightseeing locations.

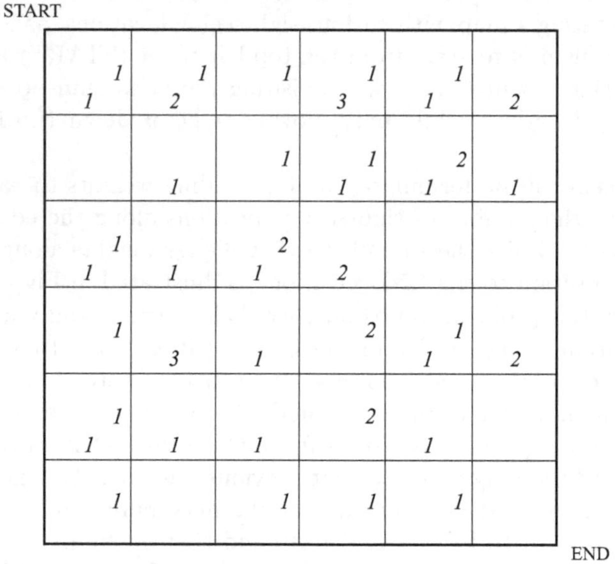

FIGURE A.2: The Manhattan Grid Problem formulated as a graph problem, where weights along each edge represent the number of sightseeing locations to potentially visit. Thus the problem is find a path from the START to the END while maximizing the sum of the edge weights.

0 and only the current edge weights are given, then recursively performing this same calculation to all consecutive positions will solve this problem.

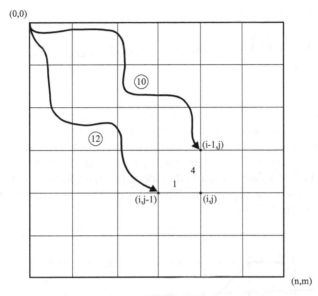

FIGURE A.3: The dynamic programming concept for determining the path to take to position (i, j) given the edge weights and the calculated maximum paths from $(0, 0)$ to the possible previous positions.

The Manhattan Grid Problem can thus be solved using the following recurrence equation where $e(w, x, y, z)$ refers to the edge weight from position (w, x) to (y, z).

$$M[i, j] = \max \begin{cases} M[i, j-1] + e(i, j-1, i, j), \\ M[i-1, j] + e(i-1, j, i, j) \end{cases}$$

This problem can be made a bit more complicated by adding diagonal edges along the grid, such as between positions $(i-1, j-1)$ to (i, j). However, the recurrence equation would only change slightly:

$$M[i, j] = \max \begin{cases} M[i, j-1] + e(i, j-1, i, j), \\ M[i-1, j] + e(i-1, j, i, j), \\ M[i-1, j-1] + e(i-1, j-1, i, j) \end{cases}$$

Now in order to link this problem to sequence alignment, let us first consider the alignment of the sequences ACTVRG and CLTRA. We can label these sequences i and j, respectively, and number each residue from the number 1 (one). Given the following alignment, therefore, we are given the path $(0,0) \rightarrow (1,0) \rightarrow (2,1) \rightarrow (2,2) \rightarrow (3,3) \rightarrow (4,3) \rightarrow (5,4) \rightarrow (6,4) \rightarrow (6,5)$.

```
1 2   3 4 5 6
A C - T V R G -
  |   |   |
- C L T - R - A
1 2 3   4   5
```

Sequence i can be placed along the top and sequence j down the left of a matrix which includes diagonal paths, as in Figure A.4. The weights of the diagonal edges are 1, and the horizontal and vertical edges are 0. Thus, we can see that the sequence alignment problem is actually the same as the Manhattan Grid Problem, where edge weights correspond to the similarity of the two residues at each position. Thus the problem is to maximize the sum of the residue similarities.

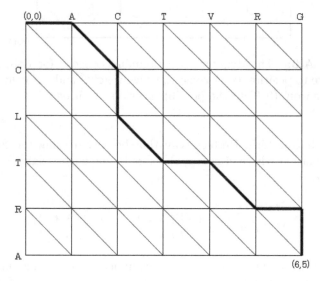

FIGURE A.4: The dynamic programming concept for aligning two sequences ACTVRG and CLTRA.

A.1.2 Sequence alignment

The two sequence alignment methods using the dynamic programming approach are the Needleman-Wunsch and Smith-Waterman algorithms. The difference between them is that the former is a global alignment algorithm whereas the latter is a local alignment algorithm.

The following is the Needleman-Wunsch algorithm (Needleman and Wunsch (1970)) comparing two sequences:

$$S[i,0] = d \cdot i,$$
$$S[0,j] = d \cdot j,$$
$$S[i,j] = \max \begin{cases} S[i,j-1] + d, \\ S[i-1,j] + d, \\ S[i-1,j-1] + w(x_i, y_j) \end{cases}$$

where $x_1 \ldots x_n$ and $y_1 \ldots y_m$ are two input sequences whose lengths are n and m, respectively, $d < 0$ is the penalty for a gap, and $w(x,y)$ denotes the score between residues x and y. Thus $S[n,m]$ gives the score of an optimal global alignment.

The Smith-Waterman algorithm (Smith and Waterman (1981b,a)) for local sequence alignments, on the other hand, differs by disallowing negative values for scores, consequently enhancing the effect of positive values, keeping matched consecutive residues ungapped and effectively terminating the alignment when the score reaches zero.

$$S[i,0] = 0,$$
$$S[0,j] = 0,$$
$$S[i,j] = \max \begin{cases} 0, \\ S[i,j-1] + d, \\ S[i-1,j] + d, \\ S[i-1,j-1] + w(x_i, y_j), \end{cases}$$

where $\max_{i,j} S[i,j]$ gives the score of an optimal local alignment.

As an example to demonstrate the difference between these two algorithms, a global and local alignment of the two sequences PROGRAMS and ROAMS will be performed. The following parameters are predefined:

$$d = -1$$
$$w(x,y) = 1 \quad \text{iff} \quad x = y, \quad 0 \quad \text{otherwise}$$

In the case of a global alignment, the results of the algorithm would be as in Table A.1. From this matrix, the resulting alignment score would be the value at position $S[n,m] = S[8,5] = 2$, and the resulting alignment would be as follows.

```
P R O G R A M S
| |     | | |
R O - - A M S
```

TABLE A.1:   Resulting table of a global alignment of two sequences.

		P	R	O	G	R	A	M	S
	0	-1	-2	-3	-4	-5	-6	-7	-8
R	-1	0	0	-1	-2	-3	-4	-5	-6
O	-2	-1	0	1	0	-1	-2	-3	-4
A	-3	-2	-1	0	1	0	0	-1	-2
M	-4	-3	-2	1	0	1	0	1	0
S	-5	-4	-3	-2	-1	0	1	0	2

TABLE A.2:   Resulting table of a local alignment of two sequences.

		P	R	O	G	R	A	M	S
	0	0	0	0	0	0	0	0	0
R	0	0	1	0	0	1	0	0	0
O	0	0	0	2	1	0	1	0	0
A	0	0	0	0	1	1	1	1	0
M	0	0	0	0	0	1	1	2	1
S	0	0	0	0	0	0	1	1	3

In contrast, for the local alignment of the same two sequences as in Table A.2, the alignment score is the maximum value in the matrix, which happens to be $S[8,5] = 3$ in this case. Backtracing this matrix, we obtain the following alignment.

```
P R O G R A M S
        | | |
        A M S
```

Note how only the longest ungapped sequence is aligned in the local sequence alignment, whereas all possible residues are matched in the global sequence alignment. The local sequence alignment consequently identifies conserved motifs.

A.2 BLOSUM (BLOcks Substitution Matrix)

The first amino acid score matrix was developed by Dayhoff et al. (1983), based on the substitution rates derived from alignments of protein sequences that are at least 85% identical. In an attempt to capture substitutions between more distantly related protein sequences, Henikoff and Henikoff (1992) developed the BLOSUM method whereby over 500 groups of related proteins were analyzed using 2000 blocks of aligned sequence segments, resulting in marked improvements in alignments and queries among these groups.

BLOSUM is computed by first deriving a frequency table from a database of so-called blocks, which are local alignments with no gaps. Thus a block could be considered as a conserved region representing a protein family. For a new member of a certain family, a set of scores for matches and mismatches is computed such that a correct alignment can be found. For each column of the block, the number of matches and mismatches of each amino acid is counted between the new sequence and every other sequence in the block. For example, if the first amino acid is A and the first column of the block contains nine A residues and one S residue, then there are nine AA matches and one AS mismatch. After repeating this count for each column in the block, all counts are summed and stored in a table. For every new sequence added to this block, the new sums are added to this table, resulting in a frequency table for all possible amino acid pairs in each column of the block.

Let f_{ij} be the total number of pairs of amino acids i, j in the frequency table. Then the observed probability of occurrence for each i, j pair can be computed as in Equation A.1.

$$q_{ij} = \frac{f_{ij}}{\sum\limits_{i=1}^{20} \sum\limits_{j=1}^{i} f_{ij}} \tag{A.1}$$

Using Equation A.1, the probability of occurrence of amino acid i in an i, j pair can be computed as in Equation A.2.

$$p_i = q_{ii} + \sum_{j \neq i} \frac{q_{ij}}{2} \tag{A.2}$$

The expected probability of occurrence of each i, j pair can then be computed as in Equation A.3

$$e_{ij} = \begin{cases} p_i p_j & \text{for} \quad i = j \\ 2 p_i p_j & \text{for} \quad i \neq j \end{cases} \tag{A.3}$$

The log odds ratio can then be computed in bit units as $s_{ij} = log_2(q_{ij}/e_{ij})$. It can thus be assumed that if the observed frequencies are as expected, $s_{ij} = 0$. If the observed frequences are less (or more) than expected, then $s_{ij} < (>) 0$.

In order to analyze the effectiveness of BLOSUM, the average mutual information, or relative entropy, H for each amino acid pair, along with the expected score E in bit units was computed as follows.

$$H = \sum_{i=1}^{20} \sum_{j=1}^{i} q_{ij} s_{ij} \tag{A.4}$$

$$E = \sum_{i=1}^{20} \sum_{j=1}^{i} p_i p_j s_{ij} \tag{A.5}$$

Now in order to reduce bias generated from the alignment of extremely similar sequences within a block, sequences are clustered within blocks such that each cluster is weighted as a single sequence. Clusters are defined by a percentage indicating percent similarity. Thus if the clustering percentage is 80%, then two sequences A and B which are $\geq$ 80% similar would be clustered, and their contributions averaged in calculating pair frequencies. For example, continuing the example above for residue A aligned to a block containing nine A and one S residue, if eight of the nine sequences with A residues are clustered, then instead of counting nine A residues, these would be counted as two A residues instead. Thus a variety of matrices can be created based on the clustering percentage. For a matrix generated based on clustering at 80% similarity, the resulting matrix is called BLOSUM 80.

Appendix B

Machine Learning Methods

This appendix chapter covers the machine learning methods of kernels and hidden Markov models (HMMs). Although there are numerous machine learning methods in the bioinformatics literature, these two have been applied to carbohydrate structures as well as glycan-related genes. Thus a brief primer on these methods is provided here.

B.1 Kernel methods and SVMs

A general overview to kernels and SVMs will be described here. For more detailed explanations, the interested reader is referred to Scholkopf and Smola (2002). Kernels can be informally defined as similarity measures that arise from a particular representation of patterns, and one of the main kernel algorithms is support vector machines (SVMs). Kernels are used to classify objects based on the parameters learned from two given classes of objects. That is, a kernel is trained on two classes of objects, and it can thus be used to classify new objects into one of the learned classes. Formally, this function can be formulated as follows:

$$(x_1, y_1), \ldots, (x_m, y_m) \in \mathcal{X} \times \{\pm 1\}$$

where $\mathcal{X}$ is some nonempty set from which the observations, or inputs x_i are taken, and y_i are the labels corresponding to x_i. Note that this is a special case where exactly two classes are given, labeled by $+1$ and -1, respectively.

Furthermore, given some new pattern $x \in \mathcal{X}$, one can predict the corresponding $y \in \{\pm 1\}$. This requires the notion of *similarity* in $\mathcal{X}$ and in $\{\pm 1\}$. Normally, the similarity between two objects x and x' may be mapped to some space of real numbers, which can be formulated as follows:

$$k : \mathcal{X} \times \mathcal{X} \to \mathbb{R}$$
$$(x, x') \mapsto k(x, x')$$

where k is the kernel function. Note that it is assumed that k is symmetric; that is, $k(x, x') = k(x', x)$ for all $x, x' \in \mathcal{X}$.

However, in general, such similarity measures are rather difficult to study. For example, it may be possible to compute the similarity of two glycans g and g' using KCaM as described in Section 4.2.1, but training based on the KCaM similarity of two classes of glycans may not necessarily enable the prediction of the classification of a new glycan simply based on this similarity score. Further information regarding the structural similarity at higher resolution, or even regarding the pathways in which the structures are involved, may be needed in order to accurately classify new structures. Indeed, it has been shown that kernels can be used to train on a variety of information (such as in Yamanishi et al. (2005)) due to the convenient forms of similarity measures that kernels can utilize, which are described next.

To start simply, a simple type of similarity measure that is particularly mathematically appealing is the dot product. For example, given two vectors $x, x' \in \mathbb{R}^N$, then the canonical dot product is defined as

$$\langle x, x' \rangle := \sum_{i=1}^{N} [x]_i [x']_i$$

where $[x]_i$ denotes the ith entry of vector x. However, in order to be able to use a dot product as a similarity measure, the input patterns must be represented as vectors in some dot product space $\mathcal{H}$, which does not necessarily need to coincide with $\mathbb{R}^N$. Thus the following mapping may be used:

$$\Phi : \mathcal{X} \to \mathcal{H}$$
$$x \mapsto x := \Phi(x).$$

In this case, the space $\mathcal{H}$ is called a feature space, and the vectorial representation of x in the feature space is represented as x. Now given two classes of features in this feature space, the fundamental idea of SVM is to construct the optimal hyperplane which separates the classes with the maximal *margin*, which refers to the distance between the hyperplane and the closest input data point. Assuming that our classes in the mapped feature space can now be linearly separated, the linear discriminant, or decision, function can be defined by linear combinations of the input vector components as in the following:

$$f(x) = \langle w, x \rangle + b$$

where $w \in \mathcal{H}$ is a weight vector and $b \in \mathbb{R}$ is a threshold. Note that $\langle w, x \rangle + b = 0$ represents the class of hyperplanes in some dot product space $\mathcal{H}$. Thus for a vector x, if $f(x) > 0$, then the model assigns this vector with a positive label; otherwise, it assigns a negative label to it.

Based on this idea, the goal, then, of finding the optimal hyperplane can be defined as the maximization of the minimum distance between vectors and the hyperplane, as follows:

$$\max_{\boldsymbol{w}\in\mathcal{H},b\in\mathbb{R}} \min\{\|\boldsymbol{x}-\boldsymbol{x}_i\| : \boldsymbol{x}\in\mathcal{H}, \langle\boldsymbol{w},\boldsymbol{x}\rangle+b=0, \quad i=1,\ldots,m\}.$$

Hence in order to construct the optimal hyperplane, the following equation needs to be solved:

$$\min_{\boldsymbol{w}\in\mathcal{H},b\in\mathbb{R}} \tau(\boldsymbol{w}) = \frac{1}{2}\|\boldsymbol{w}\|^2 \tag{B.1}$$

subject to

$$y_i(\langle\boldsymbol{w},\boldsymbol{x}_i\rangle+b) \geq c \tag{B.2}$$

for all $i=1,\ldots,m$ and any positive number $c \geq 1$. This equation can be explained by Figure B.1. The optimal hyperplane is shown as the solid line in the middle. Since these two classes are linearly separable, there exists a weight vector $\boldsymbol{w}$ and a threshold b such that $y_i(\langle\boldsymbol{w},\boldsymbol{x}_i\rangle+b) > 0$ for $i=1,\ldots,m$. $\boldsymbol{w}$ and b can be rescaled such that the point(s) closest to the hyperplane satisfy $|\langle\boldsymbol{w},\boldsymbol{x}_i\rangle+b| = 1$, by which a canonical form $(\boldsymbol{w},b)$ of the hyperplane satisfying $y_i(\langle\boldsymbol{w},\boldsymbol{x}_i\rangle+b) \geq 1$ can be obtained. Here, the margin equals $1/\|w\|$, which can be derived by the following computation.

$$\langle\boldsymbol{w},\boldsymbol{x}_1\rangle+b = 1$$
$$\langle\boldsymbol{w},\boldsymbol{x}_2\rangle+b = -1$$
$$\langle\boldsymbol{w},(\boldsymbol{x}_1-\boldsymbol{x}_2)\rangle = 2$$
$$\langle\frac{\boldsymbol{w}}{\|\boldsymbol{w}\|},(\boldsymbol{x}_1-\boldsymbol{x}_2)\rangle = \frac{2}{\|\boldsymbol{w}\|}$$

The function τ in Equation B.1 is called the objective function, and Equation B.2 represents the inequality constraints. Put together, they form a constrained optimization problem, which can be solved by using Lagrange multipliers $\alpha_i \geq 0$ and a Lagrangian L as follows:

$$L(\boldsymbol{w},b,\boldsymbol{\alpha}=(\alpha_1,\ldots,\alpha_m)) = \frac{1}{2}\|\boldsymbol{w}\|^2 - \sum_{i=1}^{m}\alpha_i(y_i(\langle\boldsymbol{x}_i,\boldsymbol{w}\rangle+b)-1). \tag{B.3}$$

The Lagrangian L must be minimized with respect to the primal variables $\boldsymbol{w}$ and b and maximized with respect to the dual variables α_i. In other words,

$$\frac{\partial}{\partial b}L(\boldsymbol{w},b,\boldsymbol{\alpha}) = 0 \quad \text{and}$$

$$\frac{\partial}{\partial \boldsymbol{w}}L(\boldsymbol{w},b,\boldsymbol{\alpha}) = 0,$$

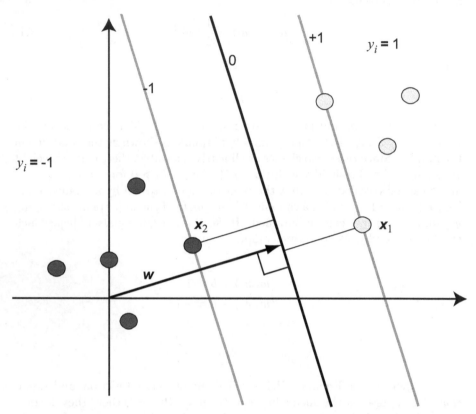

FIGURE B.1: The optimal hyperplane is shown as the solid line in the middle. Since these two classes are linearly separable, there exists a weight vector w and a threshold b such that $y_i(\langle w, x_i \rangle + b) > 0$ for $i = 1, \ldots, m$. w and b can be rescaled such that the point(s) closest to the hyperplane satisfy $|\langle w, x_i \rangle + b| = 1$, by which a canonical form (w, b) of the hyperplane satisfying $y_i(\langle w, x_i \rangle + b) \geq 1$ can be obtained. It is noted that in this case, the margin equals $1/\|w\|$.

which leads to

$$\sum_{i=1}^{m} \alpha_i y_i = 0 \tag{B.4}$$

and

$$\boldsymbol{w} = \sum_{i=1}^{m} \alpha_i y_i \boldsymbol{x}_i \tag{B.5}$$

The solution vector, the last line above, has an expansion in terms of a subset of the input patterns (those patterns with non-zero α_i), which are the support vectors (SVs). In fact, all other input examples may be ignored as they do not appear in the expansion. Thus the hyperplane is completely determined by the input patterns closest to it, regardless of the rest.

By substituting Equations B.4 and B.5 into the Lagrangian in Equation B.3, the primal variables $\boldsymbol{w}$ and b can actually be eliminated to produce the dual optimization problem, as follows:

$$\max_{\boldsymbol{\alpha} \in \mathbb{R}^m} W(\boldsymbol{\alpha}) = \sum_{i=1}^{m} \alpha_i - \frac{1}{2} \sum_{i,j=1}^{m} \alpha_i \alpha_j y_i y_j \langle \boldsymbol{x}_i, \boldsymbol{x}_j \rangle$$

with constraints $\alpha_i \geq 0$ for all $i = 1, \ldots, m$ and Equation B.4. This dual optimization problem is what is usually solved in practice.

Using the above equations, the optimal hyperplane in the feature space $\mathcal{H}$ can be found. However, there must be a way to express these formulas in terms of the actual input space $\mathcal{X}$. This is where the kernel trick can be used, as follows:

$$k(x, x') = \langle \boldsymbol{x}, \boldsymbol{x}' \rangle$$

where k is the kernel and x, x' are the input patterns. Thus replacing all the $\langle \boldsymbol{x}, \boldsymbol{x}' \rangle$ with $k(x, x')$ in the previous equations, the solution can be found based on the original input patterns. Note that the dot product is the simplest kernel, but other functions may be used instead (Jankowski and Grabczewski (2006)).

B.2 Hidden Markov models

Let us assume we are given the following problem. We are shown a set of protein sequences and asked to find a common pattern from this set that would represent this set. Conventionally, this problem would be solved by aligning the sequences and finding the consensus sequence. However, this can only be performed on highly similar sequences with little noise. We may still be able to solve this problem despite the fact that some of our given sequences may not be from the same family and some may even contain inaccurate residues.

We can reword this problem in the following manner. We are shown a set of observed sequences of symbols, which, in this case, are the set of amino acids. We are also given a set of *states*, which can output any symbol from this set based on some probability distribution. Each state can also transition to any other state based on some other probability distribution. We want to find the sequence of states and their outputs that most closely match the given set of observations. This can be done using Hidden Markov models, or HMMs (Eddy (1996)).

HMMs can be defined by a tuple $(S, \sum, \prod, A, B)$, where $S = \{s_1, s_2, \ldots, s_N\}$ is the set of states, $\sum = \{w_1, w_2, \ldots, w_M\}$ is the set of output symbols, or labels, $\prod$ is the set of initial state probabilities, A is the set of state transition probabilities, and B is the set of symbol emission probabilities. For the first time point or node in the HMM, the probability that its initial state is s_i is $\pi_i = p\{q_1 = s_i\}$. The probability that for a given node (or time point) q_{t+1} the state is s_j and it transitions from s_i is $a_{ij} = p\{q_{t+1} = s_j | q_t = s_i\}$. Finally the probability that for a given node (or time point) q_t a particular symbol w_k is outputted at state s_j is $b_{jk} = p\{o_t = w_k | q_t = s_j\}$. We make note that $a_{ij} \geq 0$ and $\sum_{j=1}^{N} a_{ij} = 1$.

Figure B.2 is an example of an HMM with four nodes, or time points, labeled $1, \ldots, 4$. Each state s_i outputs a symbol w_i, and each state transitions from one other state, except for the first node, whose starting probability uses the initial state probability. Note that the set of states used to build this HMM may be larger than four, and the outputted symbols may also vary independent of the number of nodes. Each state is also free to transition to any other state, and the objective of the HMM is to find the most likely set of states, transitions, and output symbols that match the given observations.

We note here that there are some assumptions that are made in the use of HMMs (in particular, first-order HMMs).

The Markov assumption The transition to the next state is dependent only on the current state. Note that k^{th}-order Markov models are defined as models whose states are dependent on the previous k states.

The stationarity assumption The state transition probabilities are independent of the actual time at which the transitions take place. That is, for any t_1 and t_2, $p\{q_{t_1+1} = j | q_{t_1} = i\} = p\{q_{t_2+1} = j | q_{t_2} = i\}$. In other words, the system will transition to state j with probability p_{ij}, regardless of the value of t.

The output independence assumption Regarding the observations, or the outputs, the current output is statistically independent of the previous

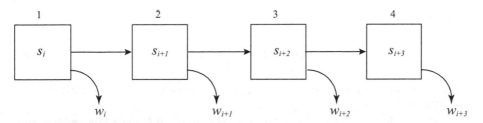

FIGURE B.2: An example of a hidden Markov model of four time points, or nodes, labeled $1, \ldots, 4$. Each state s_i outputs a symbol w_i, and each state transitions from one other state, except for the first node, whose starting probability uses the initial state probability.

outputs. Thus, given a sequence of observations $O = o_1, o_2, \ldots, o_T$, for an HMM λ, $p\{O|s_1, s_2, \ldots, s_T, \lambda\} = \sum_{t=1}^{T} p(o_t|s_t, \lambda)$.

B.2.1 The three problems of interest for HMMs

Now that we have defined HMMs, let us look at the three problems of interest when designing an HMM (Rabiner (1989)).

Problem 1: Probability evaluation What is the probability that the observations are generated by the model?

Problem 2: Optimal state sequence What is the most likely state sequence in the model that produced the observations?

Problem 3: Parameter estimation What are the most likely parameters that generate the observations most closely?

Problem 1 involves the computation of $p\{O|\lambda\}$ given our HMM and a sequence of observations $O = o_1, o_2, \ldots, o_T$. This problem can be computed efficiently using an auxiliary variable $\alpha_t(i)$, defined as the probability of observing the given sequence up to state s_i at time t.

$$\alpha_t(i) = p\{o_1, o_2, \ldots, o_t, q_t = s_i|\lambda\} \tag{B.6}$$

Using recursion, we can solve this problem by defining the initial observation as $\alpha_1(j) = \pi_j b_j(o_1)$ for $1 \le j \le N$ and using the following recursive relationship:

$$\alpha_{t+1}(j) = b_j(o_{t+1}) \sum_{i=1}^{N} \alpha_t(i) a_{ij} \tag{B.7}$$

for $1 \le j \le N$ and $1 \le t \le T-1$ where $\alpha_1(j) = \pi_j b_j(o_1)$. Using this recursion, we can compute $\alpha_T(i)$ for $1 \le i \le N$, and thus the solution to this problem can be found by

$$p\{O|\lambda\} = \sum_{i=1}^{N} \alpha_T(i). \tag{B.8}$$

This method, known as the *forward algorithm*, runs in linear time with respect to T.

For completeness, we also introduce the *backward algorithm*, which can also solve the same problem, but from the other direction. The backward probability of observing the given sequence $o_{t+1}, o_{t+2}, \ldots, o_T$ given that the current state is s_i is:

$$\beta_t(i) = p\{o_{t+1}, o_{t+2}, \ldots, o_T | q_t = i, \lambda\} \tag{B.9}$$

Similarly to the forward algorithm, the following recursive relationship holds:

$$\beta_t(i) = \sum_{j=1}^{N} \beta_{t+1}(j) a_{ij} b_j(ot+1) \tag{B.10}$$

for $1 \leq j \leq N$ and $1 \leq t \leq T - 1$ where $\beta_T(i) = 1, 1 \leq i \leq N$. Thus the solution to this problem can also be found by the backward algorithm:

$$p\{O|\lambda\} = \sum_{i=1}^{N} \beta_T(i). \tag{B.11}$$

Moreover, we can compute $p\{O|\lambda\}$ using both the forward and backward probabilities as follows:

$$p\{O|\lambda\} = \sum_{i=1}^{N} p\{O, q_t = i|\lambda\} = \sum_{i=1}^{N} \alpha_t(i)\beta_t(i) \tag{B.12}$$

Problem 2 can be solved using the Viterbi algorithm, which uses another auxiliary variable δ defined as follows:

$$\delta_t(i) = \max_{q_1,q_2,\ldots,q_{t+1}} p\{q_1, q_2, \ldots, q_{t-1}, o_1, o_2, \ldots, o_{t-1}|\lambda\} \tag{B.13}$$

This computes the highest probability for the partially observed sequence up to t and state sequence up to state i. Thus, similarly to the recursion for problem 1, using the initial probability $\delta_1(j) = \pi_j b_j(o_1)$ for $1 \leq j \leq N$, we can obtain the following recursive relationship:

$$\delta_{t+1}(j) = b_j(o_{t+1}) \max_{1 \leq i \leq N} \delta_t(i) a_{ij} \tag{B.14}$$

for $1 \leq i \leq N$ and $1 \leq t \leq T - 1$. Consequently, the solution to this problem is a matter of finding those states that provide the highest probability. In other words, we use *arg* max to obtain the most likely state sequence

$$j^* = \arg \max_{1 \leq j \leq N} \delta_T(j) \tag{B.15}$$

and then backtrack from this state to retrieve all most-likely states.

Finally, problem 3 is the actual training of the model based on the given set of observations. In general, this involves the optimization of the probabilities such that they match the set of observations. For the purposes of this book, we will focus on maximizing the total likelihood of the observations given the HMM λ, which can be described as $L_{tot} = p\{O|\lambda\}$. One of the most well-known methods to solve this problem is the EM algorithm, which is described in the next section.

B.2.2 Expectation-Maximization (EM) algorithm

EM stands for Expectation-Maximization, which is a method by which the parameters of probabilistic models can be estimated (Dempster et al. (1977)). It attempts to find the maximum likelihood of the model for a given set of observations through the iteration of two basic steps, as the name suggests: computing the expectation (E) of the likelihood and maximizing (M) the likelihood. The parameters found in the M step are subsequently used to compute the E step, and the process repeats until some condition is reached.

B.2.2.1 Baum-Welch algorithm

One of the popular EM algorithms for HMMs is the Baum-Welch algorithm (Baum et al. (1970)), which is also known as the forward-backward algorithm since it uses the forward and backward probabilities described in Section B.2.1. In addition, two other auxiliary probabilities are used, ξ and γ, defined as follows:

$$\xi_t(i,j) = p\{q_t = i, q_{t+1} = j|O, \lambda\} \tag{B.16}$$

$$\gamma_t(i) = p\{q_t = i|O, \lambda\} \tag{B.17}$$

Note that these two probabilities are related and as such can be defined as

$$\gamma_t(i) = \sum_{j=1}^{N} \xi_t(i,j) \tag{B.18}$$

for $1 \leq i \leq N$ and $1 \leq t \leq M$.

Here, $\xi_t(i,j)$ is the probability of being in state i at time t and in state j at time $t+1$. Using the definition of conditional probability, we can rewrite $\xi_t(i,j)$ as

$$\xi_t(i,j) = \frac{p\{q_t = i, q_{t+1} = j, O|\lambda\}}{p\{O|\lambda\}}. \tag{B.19}$$

Referring back to Equation D.12 we see that we can rewrite this again using the forward and backward probabilities as

$$\xi_t(i,j) = \frac{\alpha_t(i)a_{ij}\beta_{t+1}(j)b_j(o_{t+1})}{\sum_{i=1}^{N}\sum_{j=1}^{N}\alpha_t(i)a_{ij}\beta_{t+1}(j)b_j(o_{t+1})}. \tag{B.20}$$

For $\gamma_t(i)$, we can also use the definition of conditional probability to get

$$\gamma_t(i) = \frac{p\{q_t = i, O|\lambda\}}{p\{O|\lambda\}} \tag{B.21}$$

which can be rewritten using Equation B.12 as

$$\gamma_t(i) = \frac{\alpha_t(i)\beta_t(i)}{\sum_{i=1}^{N} \alpha_t(i)\beta_t(i)} \tag{B.22}$$

With these probabilities defined, we can now describe the EM algorithm.

B.2.2.1.1   Expectation step   Assuming we have our HMM λ we simply calculate all α's and β's using the recursions B.7 and B.10. Next, we compute the expectation values ξ and γ using Equations B.20 and B.18.

B.2.2.1.2   Maximization step   The maximization step involves the re-estimation of our original parameters in an attempt to increase the maximum likelihood. Thus the following re-estimation formulas are used.

$$\bar{\pi} = \gamma_1(i) \qquad\qquad \text{for}\quad 1 \le i \le N$$

$$\bar{a}_{ij} = \frac{\sum_{t=1}^{T-1} \xi_t(i,j)}{\sum_{t=1}^{T-1} \gamma_t(i)} \qquad\qquad \text{for}\quad 1 \le i \le N, 1 \le j \le N$$

$$\bar{b}_j(k) = \frac{\sum_{t=1}^{T} \gamma_t(j) \quad \text{for}\quad o_t = v_k}{\sum_{t=1}^{T} \gamma_t(j)} \qquad \text{for}\quad 1 \le j \le N, 1 \le k \le M$$

The likelihood for the model given the observations is computed using any of the equations for solving Problem 1: Equations B.8, B.11 or B.12. These updated parameters are used in another round of the E and M steps, and the likelihood is computed once more. If the new likelihood differs from the previous likelihood by a margin within a set threshold, then the recursion stops. It has been proven that the EM algorithm always converges.

B.2.3   Hidden tree Markov models

The hidden tree Markov model (HTMM) was first developed for wavelet-based statistical signal processing by which layers of connected HMMs were trained (Crouse et al. (1998)). This model was independently developed by another group some time later for document image classification, by which documents of images could be automatically classified (Diligenti et al. (2003)). This model was based on the notion that hidden Markov models can be considered as a special case of Bayesian networks (Smyth et al. (1997)) and was thus one of the first extension of HMMs as a possible architecture for connectionist learning of data structures.

To define the model, the input is denoted as a tree Y, with the set of vertices V. The model then is composed of the set of states $S = \{s_1, s_2, \ldots, s_N\}$, the set of output symbols $\sum = \{w_1, w_2, \ldots, w_M\}$, the set of initial probabilities $\prod$, the set of state transition probabilities A and the set of symbol emission probabilities B. An example illustrating the HTMM is shown in Figure B.3. In this figure, seven nodes labeled $1, \ldots, 7$ correspond to seven states $s_1, \ldots, s_7$, with each state outputting a symbol w_i. Each state transitions from one other state, its parent, except for the first root node, whose starting probability uses the initial state probability.

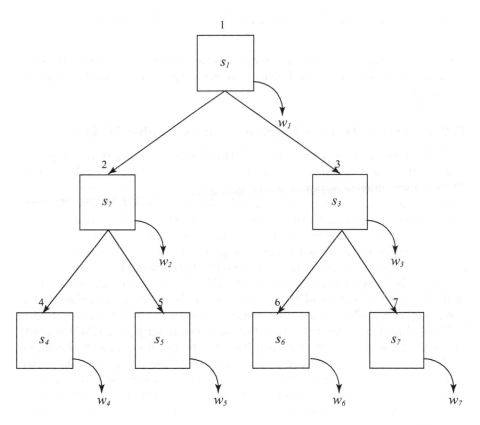

FIGURE B.3: An example of a hidden tree Markov model of seven nodes, labeled $1, \ldots, 7$. Each state s_i outputs a symbol w_i, and each state transitions from one other state, its parent, except for the first node, whose starting probability uses the initial state probability.

The EM algorithm for training HTMM was developed similar to that of the HMM algorithm, except that instead of a forward-backward algorithm, an

TABLE B.1:   Parameters for profile HMM.

$a_{M_j M_{j+1}}$	transition between match states at position j
$a_{M_j I_{j+1}}$	transition from match state at position j to insert state
$a_{M_j X_{j+1}}$	transition from match state at position j to delete state
$a_{I_j I_j}$	transition from insert state at position j back to itself
$a_{I_j M_{j+1}}$	transition from insert state at position j to match state
$a_{I_j X_{j+1}}$	transition from insert state at position j to delete state
$a_{X_j X_{j+1}}$	transition between delete states at position j
$a_{X_j I_{j+1}}$	transition from delete state at position j to insert state
$a_{X_j M_{j+1}}$	transition from delete state at position j to match state
$a_{X_j M_{j+1}}$	transition from delete state at position j to match state
b_{jk}	output probability of symbol w_k at state s_j

upward-downward algorithm was used. Interested readers are referred to the original papers for the detailed algorithms since they are beyond the scope of this book.

B.2.4   Profile Hidden Markov models (profile HMMs)

Profile HMMs are a special type of HMM that was initially developed to align multiple protein sequences efficiently (Krogh et al. (1994); Eddy (1998)). Thus it uses different types of states to represent gaps and alignments. For each consensus column of a given multiple alignment, a *match* state M_i stores the distribution of residues in column i. An *insert* state I_i at column i represent one or more inserted residues between columns i and $i + 1$. A *delete* state X_i represents a deleted residue, or gap at column i. Consequently, the number of states now depends on the length of the given alignment.

The probability parameters for profile HMMs are now restricted to pairs of distinct states because of the positioning of the columns according to the given alignment. Thus we are now given the parameters as listed in Table B.1.

The estimation of the state transition parameters for profile HMM consists of counting the frequency of a transition in the given alignment. Emission probabilities are estimated using the frequency of the emission in the alignment. Furthermore, to account for those transitions or emissions that are not present in the alignment, pseudocounts are usually introduced such that no probability is zero. Thus the estimation procedure consists of the following two equations:

$$a_{kl} = \frac{A_{kl} + r_{kl}^a}{\sum_{q \in S}(A_{kq} + r_{kq}^a)} \tag{B.23}$$

$$b_{jk} = \frac{E_{jk} + r_{jk}^e}{\sum_{w \in \Sigma}(E_{jw} + r_{jw}^e)} \tag{B.24}$$

where A_{kl} is the number of transitions from state k to state l, E_{jk} is the number of times that symbol w_k was output from state j, Q is the set of possible states, r^a_{jk} is a pseudocount for A_{jk} and $r^e_j k$ is a pseudocount for e_{jk}.

Once a model is trained on a number of sequences from a family, a score can be generated to assess how well a new protein sequence fits with the family, as was possible with HMM. However, in addition, a multiple alignment can now be generated, from which a consensus sequence can be extracted. The Baum-Welch procedure can be used to train profile HMMs in order to build profiles from the unaligned sequences.

The profile HMM has been implemented as software called HMMER, which has been used to generate a database of protein sequence profiles. This database is called Pfam (Finn et al. (2008)).

Appendix C

Glycomics Technologies

Glycomics technologies that generate carbohydrate data are covered in this appendix chapter. In order to fully appreciate the automatic annotation methods for mass spectra and the NMR data for carbohydrates, a brief primer to these methods are provided here.

C.1 Mass spectrometry (MS)

Mass spectrometry is an analytical tool for measuring the molecular mass of a sample. Structural information can be obtained using higher resolution techniques such as by using multiple analyzers, known as tandem MS. In order to systematize the carbohydrate fragmentation patterns, Domon and Costello (1988) proposed a systematic nomenclature consisting of labels A, B, C and X, Y, Z. These labels are subscripted with the position relative to the termini and superscripted with the cleavages within carbohydrate rings. Figure C.1 illustrates this nomenclature, which is currently widely used.

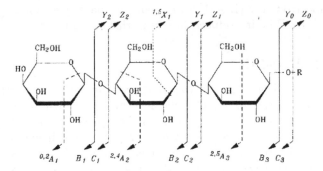

FIGURE C.1: The Domon-Costello nomenclature for carbohydrate fragmentation patterns. *Figure reused by permission of Springer.*

In addition to MS, glycosidic linkages may also be identified by permethylation analysis of the sugars, where the free hydroxyl groups on the carbohydrate sample are initially converted to methyl esters such that OH becomes OMe. Complete methylation is called permethylation, and permethylated substances can be analyzed by gas chromatography-mass spectrometry (GC-MS) to perform linkage analysis (Price (2008)).

Mass spectrometers consist of three fundamental parts: the ionization source, the analyzer, and the detector. The sample is first introduced into the ionization source, where the sample molecules are ionized (which are easier to manipulate than neutral molecules). These ions are extracted into the analyzer, which separates the molecules according to their mass-to-charge ratios (m/z). The separated ions are detected, and this signal is sent to a data system where the m/z ratios are stored together with their relative abundance as a m/z spectrum.

These three parts of the mass spectrometer are often maintained under high vacuum such that the ions have a reasonable chance of traveling from one end of the instrument to the other without being hindered by air molecules (Keen and Ashcroft (1999); Ashcroft (1997)).

For large molecular weight biomolecules, it is important to note that the measured mass is actually the average mass and that the peak envelope extends over many individual masses, corresponding to the peaks of isotopes of the molecule. As larger molecules are analyzed, wider peak envelopes will be recorded, thus obscuring the actual peaks at larger mass values.

Sample ionization can be performed in a number of ways. However, explaining them all is beyond the scope of this book. The methods by which the algorithms introduced in Section 4.3.4 were developed will be described here. Thus, the basic principles of the matrix-assisted laser desorption ionization mass spectrometry (MALDI-MS), FT-ICR (Fourier Transform-Ion Cyclotron Resonance) mass spectrometry with SORI-CID (Sustained Off-Resonance Irradiation Collision-Induced Dissociation) fragmentation, liquid chromatography mass spectrometry (LC-MS), and tandem mass spectrometry (MS/MS) methodologies will be introduced briefly in this section.

C.1.1   MALDI-MS

In a MALDI-MS experiment, the biological sample is co-crystallized on a metal plate with a vast excess of a chemical matrix, which has low molecular weight and can absorb UV. The plate is then placed into a vacuum where the crystal is irradiated with intense laser pulses which excite the matrix molecules, leading to the dislodging, or sputtering, of the ions of the sample and matrix from the surface of the crystal. The matrix used here may depend on the type of sample being analyzed, such as 2,5-dihydroxybenzoic acid (DHB) for native and derivatized glycans, or α-cyano-4-hydroxycinnamic acid (CHCA) for glycopeptides. The resulting ions are singly charged regardless of molecular weight, with minimal fragmentation. Thus it has been used extensively in the mass profiling of mixtures of glycans (Dell et al. (2007)).

TOF (time-of-flight) is often used as the analyzer with MALDI, where ions are formed in pulses, and the time it takes for the ions to reach the detector is measured, assuming that the smaller ions reach the detector faster than larger ions. The mass-to-charge ratio of an ion is then computed using its drift time squared, as follows:

$$\frac{m}{z} = \frac{2t^2 K}{L^2}$$

where t is the drift time, L is the drift length, m is the mass, K is the kinetic energy of the ion, and z is the number of charges on the ion.

C.1.2 FT-ICR

Fourier transform-ion cyclotron resonance (FT-ICR) mass spectrometry computes the mass-to-charge ratio of ions based on the cyclotron frequency of the ions in a fixed magnetic field. The ion cyclotron frequency, radius, velocity, and energy as a function of ion mass, ion charge and magnetic field strength can be modeled directly from the motion of an ion in a spatially uniform static magnetic field. This ion cyclotron motion can be spatially observed by applying a spatially uniform radio frequency (RF) electric field to excite the ions at the same frequency (or resonance) as the ion cyclotron frequency. This ion cyclotron resonance (ICR) signal results from the detection of an oscillating "image" charge on two conductive infinitely extended opposed parallel electrodes. A frequency-domain spectrum, which is convertible to a mass-domain spectrum, is obtained by Fourier transformation of the digitized time-domain ICR signal. Thus FT-ICR MS may be performed in essentially the same way as ion traps of other shapes, such as cylinders. Furthermore, the collisions broaden the ICR signal in a simple way, making it actually possible to cool and compress the ion packets for improved detection and multiple remeasurement. It is thus possible to perform tandem-in-time mass spectrometry: MS/MS or even MS^n (Marshall et al. (1998)).

In an FT-ICR instrument, ion activation can be performed by collisions of ions with neutrals in what is called collision-induced dissociation (CID) or multiphoton infrared photodissociation (IRMPD), among others. The most popular method for CID of macromolecular ions is sustained off-resonance irradiation (SORI). During SORI-CID, ions are excited by the application of an RF electric field pulse with a frequency "off-resonance" with the ion's natural cyclotron frequency, ω_c. This frequency pulse results in a maximal translational energy E_{tr} given by $E_{tr} = (E/\sqrt{2})^2 e^2 / [2m(\omega - \omega_c)^2] \sin^2(\omega - \omega_c) t/2$, where E is the amplitude of the RF pulse, e is the electric charge, ω is the excitation frequency, and t is the duration of the RF pulse. The "off-resonance" pulse causes the ion to undergo some number n of acceleration-deceleration cycles given by $n = t(\omega - \omega_c)$. These cycles cause the ion to be confined in the cell for a sustained period of irradiation ($> 500ms$). In the presence of a low-mass gas target such as N_2 or Ar at a pressure of approximately 10^{-6} torr, SORI produces many sequential, low-energy inelastic collisions, which slowly activate the molecules, resulting in dissociation occurring mainly through the lowest energy fragmentation channels (Herrmann et al. (2005)).

The photodissociation of large biomolecules (IRMPD) has also proven effective. Typically, IR (10.6μm) laser photons are used for "slow heating" and fragments similar to those obtained by CID are produced. One advantage of IRMPD is that gas pulses are unnecessary. Thus high-resolution FT-ICR detection can be obtained quickly after dissociation (Marshall et al. (1998)).

C.1.3   LC-MS (HPLC)

Liquid chromatography-mass spectrometry (LC-MS), or high-performance liquid chromatography (HPLC), is a highly powerful technique for the identification of chemicals in a mixture. For beginners, chromatography is a technique for the separation of mixtures, involving passing the mixture sample dissolved in a "mobile phase" through a "stationary phase," which separates the target analyte from other molecules in the mixture to isolate it. In liquid chromatography (LC), the mobile phase is a liquid. In HPLC, using a liquid at high pressure (mobile phase), the sample is pumped through a column that is packed with either very small particles or a porous monolithic layer (the stationary phase). HPLC may be performed in normal phase or reversed phase. Normal phase is when the stationary phase is more polar compared to the mobile phase, whereas the opposite is reversed phase. Thus the differences in polarity of the chemicals in the mixture cause the compounds to be retained for different periods, consequently allowing separation. Combined with mass spectrometry, in LC-MS, the chemicals separated by LC are analyzed using MS, or even tandem MS (Tomer et al. (1994)).

C.1.4   Tandem MS

Tandem MS (MS/MS or MS^n) consists of two or more MS rounds, each more refined than the previous. This enables the capture of more detailed information regarding the structure of the analyte. MS-MS analyses usually consist of more than one analyzer, or use an ion trap where individual ions can be uniquely selected. Thus tandem MS consists of the fragmentation of the sample inside the instrument and further analysis of the generated products. That is, after the activation of primary, or precursor, ions, dissociation or reaction, mass analysis of the resulting secondary, or product, ions according to the quotient mass-per-charge ratio is performed. In MS^n, n rounds of fragment selection and further fragmentation is performed (Keen and Ashcroft (1999)).

C.2   Nuclear magnetic resonance (NMR)

NMR exploits the magnetic properties of nuclei, which can be detected to identify the molecules in a sample. NMR spectroscopy is the use of NMR to study for example the chemical structure of a sample.

The neutrons and protons composing the atomic nucleus have the intrinsic quantum property of spin, which comes in multiples of 1/2 and can be positive (+) or negative (-). Thus the overal spin of the nucleus is determined by the spin quantum number, which depends on the number of protons and neutrons in a given isotope. If these are even, then the spin quantum number is zero. If the number of neutrons plus the number of protons is odd, then the nucleus has a half-integer spin. On the other hand, if the number of neutrons and the number of protons are both odd, then the nucleus has an integer spin (e.g., 1, 2, 3). In NMR, the unpaired nuclear spins where the total spin number is not zero is of importance.

NMR consists of two steps. First, the magnetic nuclear spins are aligned (or polarized) in an applied, constant magnetic field. Second, this alignment of the nuclear spins is perturbed by radio frequency (RF) photon pulses, causing the spins to transition between different energy states. The signal in NMR spectroscopy results from the difference between the energy absorbed by the spins which transition from the lower to higher energy states, and the energy emitted by the spins which simultaneously transition from the higher to lower energy states. This resonance, or exchange of energy at a specific frequency between the spins and the spectrometer, is what gives NMR its sensitivity.

Chemically different protons have different electronic environments, and it is these differences that cause the protons to behave differently in the applied magnetic fields. In order to standardize the NMR scale, tetramethylsilane (TMS) was selected as the 0 reference point to which all protons can be compared. Since TMS generates only one signal, all NMR signals of a sample are now referenced to this single signal, which is referred to as the chemical shift, measured in parts per million, or ppm (Friebolin (2005)).

References

Angata, T. and E. B.-V. der Linden (2002). I-type lectins. *Biochim. Biophys. Acta 1572*(2-3), 294–316.

Aoki, K., H. Mamitsuka, T. Akutsu, and M. Kanehisa (2005). A score matrix to reveal the hidden links in glycans. *Bioinformatics 21*(8), 1457–1463.

Aoki, K., N. Ueda, A. Yamaguchi, M. Kanehisa, T. Akutsu, and H. Mamitsuka (2004). Application of a new probabilistic model for recognizing complex patterns in glycans. In *Proc. 12th ISMB*.

Aoki, K., A. Yamaguchi, Y. Okuno, T. Akutsu, N. Ueda, M. Kanehisa, and H. Mamitsuka (2003). Efficient tree-matching methods for accurate carbohydrate database queries. *Genome Informatics 14*, 134–143.

Aoki, K., A. Yamaguchi, N. Ueda, T. Akutsu, H. Mamitsuka, S. Goto, and M. Kanehisa (2004). KCaM (KEGG Carbohydrate Matcher): a software tool for analyzing the structures of carbohydrate sugar chains. *Nucl. Acids Res. 32*, W267–W272.

Ashcroft, A. (1997). *Ionization methods in organic mass spectrometry*. Cambridge: Royal Society of Chemistry.

Bach, F., R. Thibaux, and M. Jordan (2005). Computing regularization paths for learning multiple kernels. *Adv. Neural Inform. Process. Syst. 17*, 73–80.

Banin, E., Y. Neuberger, Y. Altshuler, A. Halevi, O. Inbar, D. Nir, and A. Dukler (2002). A novel Linear Code(R) nomenclature for complex carbohydrates. *Trends in Glycoscience and Glycotechnology 14*(77), 127–137.

Baum, L., T. Petrie, G. Soules, and N. Weiss (1970). A maximization technique occurring in the statistical analysis of probabilistic functions of Markov chains. *Ann. Math. Stat. 41*(1), 164–171.

Beisel, H.-G., S. Kawabata, S. Iwanaga, R. Huber, and W. Bode (1999). Tachylectin-2: crystal structure of a specific GlcNAc/GalNAc-binding lectin involved in the innate immunity host defense of the Japanese horseshoe crab Tachypleus tridentatus. *EMBO J. 18*, 2313 – 2322.

Bern, M. and D. Goldberg (2005). EigenMS: de novo analysis of peptide tandem mass spectra by spectral graph partitioning. In S. Miyano et al. (Eds.), *Ninth International Conference on Research in Computational Molecular Biology (RECOMB)*, pp. 357–372.

Biessen, E., F. Noorman, M. van Teijlingen, J. Kuiper, M. Barrett-Bergshoeff, M. Bijsterbosch, D. Rijken, and T. van Berkel (1996). Lysine-based cluster mannosides that inhibit ligand binding to the human mannose receptor at nanomolar concentration. *J. Biol. Chem. 271*, 28024–28030.

Bigg, H., R. Wait, A. Rawan, and T. Cawston (2006). The mammalian chitinase-like lectin, YKL-40, binds specifically to type I collagen and modulates the rate of type I collagen fibril formation. *J. Biol. Chem. 281*, 21082–21095.

Bochner, B., R. Alvarez, P. Mehta, N. Bovin, O. Blixt, J. White, and R. Schnaar (2005). Glycan array screening reveals a candidate ligand for siglec-8. *J. Biol. Chem. 280*(6), 4307–4312.

Bohne, A., E. Lang, and C.-W. von der Lieth (1999). SWEET - WWW-based rapid 3D construction of oligo- and polysaccharides. *Bioinformatics 15*, 767–768.

Bohne-Lang, A., E. Lang, T. Forster, and C.-W. von der Lieth (2001). LINUCS: LInear Notation for Unique description of Carbohydrate Sequences. *Carbohydrate Research 336*(1), 1–11.

Bohne-Lang, A. and C.-W. von der Lieth (2005). GlyProt: in silico glycosylation of proteins. *Nucleic Acids Research 33*, W214–W219.

Breton, C., L. Snajdrova, C. Jeanneau, J. Koca, and A. Imberty (2006). Structures and mechanisms of glycosyltransferases. *Glycobiology 16*(2), 29R–36R.

Brewer, C. (1996). Multivalent lectin-carbohydrate cross-linking interactions. *Chemtracts - Biochem. Mol. Biol. 6*, 165–179.

Bucior, I. and M. Burger (2004). Carbohydrate-carbohydrate interactions in cell recognition. *Curr. Opin. Struct. Biol. 14*, 631–637.

Cammarata, M., G. Benenati, E. Odom, G. Salerno, A. Vizzini, G. Vasta, and N. Parrinello (2007). Isolation and characterization of a fish F-type lectin from gilt head bream (Sparus aurata) serum. *Biochimica et Biophysica Acta 1770*, 150–155.

Campbell, J., G. Davies, V. Bulone, and B. Henrissat (1997). A classification of nucleotide-diphospho-sugar glycosyltransferases based on amino acid sequence similarities. *Biochem. J. 326*, 929–939.

Cantarel, B., P. Coutinho, C. Rancurel, T. Bernard, V. Lombard, and B. Henrissat (2009). The Carbohydrate-Active EnZymes database (CAZy): an expert resource for glycogenomics. *Nucleic Acids Res. 37*, D233–D238.

Ceroni, A., A. Dell, and S. Haslam (2007). The GlycanBuilder: a fast, intuitive and flexible software tool for building and displaying glycan structures.

Source Code for Biology and Medicine 2, 3.

Ceroni, A., K. Maass, H. Geyer, R. Geyer, A. Dell, and S. Haslam (2008). GlycoWorkBench: a tool for the computer-assisted annotation of mass spectra of glycans. *Journal of Proteome Research 7*(4), 1650–1659.

Cooper, C., E. Gasteiger, and N. Packer (2001). GlycoMod – a software tool for determining glycosylation compositions from mass spectrometric data. *Proteomics 1*(2), 340–349.

Crouse, M., R. Nowak, and R. Baraniuk (1998). Wavelet-based statistical signal processing using hidden Markov models. *IEEE Trans. on Sig. Proc. 46*, 886–902.

Dahms, N. and M. Hancock (2002). P-type lectins. *Biochim. Biophys. Acta 1572*(2-3), 317–340.

Dayhoff, M., W. Barker, and L. Hunt (1983). Establishing homologies in protein sequences. *Methods in Enzymology 91*, 524.

Dell, A., S. Chalabi, P. Hitchen, J. Jang-Lee, V. Ledger, S. North, P.-C. Pang, S. Parry, M. Sutton-Smith, B. Tissot, H. Morris, M. Panico, and S. Haslam (2007). *Comprehensive glycoscience*, Chapter 2.02, pp. 69–100. Oxford: Elsevier Ltd.

Dempster, A., N. Laird, and D. Rubin (1977). Maximum likelihood from incomplete data via the EM algorithm. *J. R. Statist. Soc. B 39*, 1–38.

Diligenti, M., P. Frasconi, and M. Gori (2003). Hidden tree Markov models for document image classification. *IEEE Trans. on PAMI 25*(4), 519–523.

Domon, B. and C. Costello (1988). A systematic nomenclature for carbohydrate fragmentations in FAB-MS/MS spectra of glycoconjugates. *Glycoconjugate 5*, 397–409.

Doubet, S., K. Bock, D. Smith, A. Darvill, and P. Albersheim (1989). The complex carbohydrate structure database. *Trends Biochem. Sci. 14*, 475–477.

Drickamer, K. (1988). Two distinct classes of carbohydrate-recognition domains in animal lectins. *J. Biol. Chem. 263*, 9557–9560.

Eddy, S. (1996). Hidden Markov models. *Current Opinion in Structural Biology 6*, 361–365.

Eddy, S. (1998). Profile hidden Markov models. *Bioinformatics 14*(9), 755–763.

Elsner, M., H. Hashimoto, and T. Nilsson (2003). Cisternal maturation and vesicle transport: join the band wagon! *Mol. Membr. Biol. 20*, 221–229.

Endo, Y., M. Matsushita, and T. Fujita (2007). Role of ficolin in innate immunity and its molecular basis. *Immunobiology 212*(4-5), 371–379.

Finn, R., J. Tate, J. Mistry, P. Coggill, J. Sammut, H. Hotz, G. Ceric, K. Forslund, S. Eddy, E. Sonnhammer, and A. Bateman (2008). The Pfam protein families database. *Nucleic Acids Research 36*, D281–D288.

Fischler, M. and R. Bolles (1981). Random sample concensus: a paradigm for model fitting with applications to image analysis and automated cartography. *Comm. ACM 24*(6), 381–395.

Frank, M., P. Gutbrod, C. Hassayoun, and C.-W. von der Lieth (2003). Dynamic molecules: molecular dynamics for everyone. An internet-based access to molecular dynamic simulations: basic concepts. *J. Mol. Model. 9*, 308–315.

Frank, M., T. Lutteke, and C.-W. von der Lieth (2007). GlycoMapsDB: a database of the accessible conformational space of glycosidic linkages. *Nucleic Acids Research 35*, D287–D290.

Freeze, H. (2006). Genetic defects in the human glycome. *Nature Reviews 7*, 537–551.

Friebolin, H. (2005). *Basic One- and Two-Dimensional NMR spectroscopy*. Wiley-VCH.

Gabius, H.-J. (2008). Glycans: bioactive signals decoded by lectins. *Biochem. Soc. Trans. 36*, 1491–1496.

Garner, O. and L. Baum (2008). Galectin-glycan lattices regulate cell-surface glycoprotein organization and signalling. *Biochem. Soc. Trans. 36*, 1472–1477.

Garred, P. (2008). Mannose-binding lectin genetics: from A to Z. *Biochemical Society Transactions 36*, 1461–1466.

Glenn, K., R. Nelson, H. Wen, A. Mallinger, and H. Paulson (2008). Diversity in tissue expression, substrate binding, and SCF complex formation for a lectin family of ubiquitin ligases. *J. Biol. Chem. 283*(19), 12717–12729.

Gokudan, S., T. Muta, R. Tsuda, K. Koori, T. Kawahara, N. Seki, Y. Mizunoe, S. Wai, S. Iwanaga, and S. Kawabata (1999). Horseshoe crab acetyl group-recognizing lectins involved in innate immunity are structurally related to fibrinogen. *Proc. Natl. Acad. Sci. USA 96*, 10086–10091.

Goldberg, D., M. Bern, B. Li, and C. Lebrilla (2006). Automatic determination of *O*-glycan structure from fragmentation spectra. *Journal of Proteome Research 5*, 1429–1434.

Goldberg, D., M. Bern, S. Parry, M. Sutton-Smith, M. Panico, H. Morris, and A. Dell (2007). Automated *N*-glycopeptide identification using a combination of single- and tandem-ms. *Journal of Proteome Research 6*, 3995–4005.

Goldberg, D., M. Sutton-Smith, J. Paulson, and A. Dell (2005). Automatic annotation of matrix-assisted laser desorption/ionization *N*-glycan spectra. *Proteomics 5*, 865–875.

Hakomori, S. and Y. Igarashi (1995). Functional role of glycosphingolipids in cell recognition and signaling. *J. Biochem. 118*, 1091–1103.

Hand, D. and R. Till (2001). A simple generalisation of the area under the ROC curve for multiple classification problems. *Machine Learning 45*, 171–186.

Hanley, J. and B. McNeil (1982). The meaning and use of the area under a receiver operating characteristic (ROC) curve. *Radiology 143*, 29–36.

Hashimoto, K., K. Aoki-Kinoshita, N. Ueda, M. Kanehisa, and H. Mamitsuka (2006). A new efficient probabilistic model for mining labeled ordered trees. In *Proc. KDD*, pp. 177–186.

Hashimoto, K., S. Goto, S. Kawano, K. Aoki-Kinoshita, N. Ueda, M. Hamajima, T. Kawasaki, and M. Kanehisa (2006). KEGG as a glycome informatics resource. *Glycobiology 16*(5), 63R–70R.

Hashimoto, K., S. Kawano, S. Goto, K. Aoki-Kinoshita, M. Kawasima, and M. Kanehisa (2005). A global representation of the carbohydrate structures: a tool for the analysis of glycan. *Genome Informatics 16*(1), 214–222.

Hashimoto, K., I. Takigawa, M. Shiga, M. Kanehisa, and H. Mamitsuka (2008). Mining significant tree patterns in carbohydrate sugar chains. In *Proc. 7th ECCB*.

Hebert, D., S. Garman, and M. Molinari (2005). The glycan code of the endoplasmic reticulum: asparagine-linked carbohydrates as protein maturation and quality-control tags. *Trends in Cell Biology 15*(7), 364–370.

Henikoff, S. and J. G. Henikoff (1992). Amino acid substitution matrices from protein blocks. *Proc. Natl. Acad. Sci. 89*(22), 10915–10919.

Herget, S., R. Ranzinger, K. Maass, and C.-W. von der Lieth (2008). GlycoCT – a unifying sequence format for carbohydrates. *Carbohydrate Research 343*, 2162–2171.

Herrmann, K., A. Somogyi, V. Wysocki, L. Drahos, and K. Vekey (2005). Combination of sustained off-resonance irradiation and on-resonance excitation in FT-ICR. *Anal. Chem. 77*, 7626–7638.

Hirabayashi, J., T. Hashidate, Y. Arata, N. Nishi, T. Nakamura, M. Hirashima, T. Urashima, T. Oka, M. Futai, W. Muller, F. Yagi, and K. Kasai (2002). Oligosaccharide specificity of galectins: a search by frontal affinity chromatography. *Biochim Biophys Acta 1572*(2-3), 232–254.

Hizukuri, Y., Y. Yamanishi, O. Nakamura, F. Yagi, S. Goto, and M. Kanehisa (2005). Extraction of leukemia specific glycan motifs in humans by computational glycomics. *Carbohydr. Res. 340*, 2270–2278.

Hokama, A., E. Mizoguchi, and A. Mizoguchi (2008). Roles of galectins in inflammatory bowel disease. *World J. Gastroenterol. 14*(33), 5133–5137.

Holmskov, U., S. Thiel, and J. Jensenius (2003). Collectins and ficolins: humoral lectins of the innate immune defense. *Annu. Rev. Immunol. 21*, 547–578.

Honda, S., M. Kashiwagi, K. Miyamoto, Y. Takei, and S. Hirose (2000). Multiplicity, structures, and endocrine and exocrine natures of eel fucose-binding lectins. *J. Biol. Chem. 275*(42), 33151–33157.

Horan, N., L. Yan, H. Isobe, G. Whitesides, and D. Kahane (1999). Nonstatistical binding of a protein to clustered carbohydrates. *Proc. Natl. Acad. Sci. USA 96*, 11782–11786.

Hosokawa, N., I. Wada, Y. Natsuka, and K. Nagata (2006). EDEM accelerates ERAD by preventing aberrant dimer formation of misfolded alpha1-antitrypsin. *Genes to Cells 11*(5), 465–476.

Hossler, P., L.-T. Goh, M. Lee, and W.-S. Hu (2006). GlycoVis: visualizing glycan distribution in the protein *N*-glycosylation pathway in mammalian cells. *Biotechnology and Bioengineering 95*(5), 946–960.

Hossler, P., B. Mulukutla, and W.-S. Hu (2007). Systems analysis of *N*-glycan processing in mammalian cells. *PLoS ONE 8*, e713.

Hudson, S., N. Bovin, P. Crocker, and B. Bochner (2009). Polymers containing 6'-sulfated sialyl Lewis x (6'-su-sLex) selectively engage Siglec-8 on human eosinophils. *Journal of Allergy and Clinical Immunology 123*(2), S269.

Imberty, A., M. Delage, Y. Bourne, C. Cambillau, and S. Perez (1991). Data bank of three-dimensional structures of disaccharides: Part II, N-acetyllactosaminic type *N*-glycans. Comparison with the crystal structure of a biantennary octasaccharide. *Glycoconjugate J. 8*, 456–483.

Imberty, A., S. Gerber, V. Tran, and S. Perez (1990). Data bank of three-dimensional structures of disaccharides. A tool to build 3D structures of oligosaccharides. Part I. Oligo-mannose type *N*-glycans. *Glycoconjugate J. 7*, 37–54.

Ishino, T., T. Kunieda, S. Natori, K. Sekimizu, and T. Kubo (2007). Identification of novel members of the Xenopus Ca2+-dependent lectin family and analysis of their gene expression during tail regeneration and development. *J. Biochem. 141*(4), 479–488.

Itai, S., S. Arii, R. Tobe, A. Kitahara, Y.-C. Kim, H. Yamabe, H. Ohtsuki, Y. Kirihata, K. Shigeta, and R. Kannagi (1988). Significance of 2-3 and 2-6 sialylation of Lewis A antigen in pancreas cancer. *Cancer 61*, 775–787.

Itonori, S. and M. Sugita (2005). Diversity of oligosaccharide structures of glycosphingolipids in invertebrates. *Trends in Glycoscience and Glycotechnology 17*(93), 15–25.

Iwama, M., Y. Ogawa, K. Ohgi, T. Tsuji, and M. Irie (2001). Enzymatic properties of sialic acid binding lectin from Rana catesbeiana modified with a water-soluble carbodiimide in the presence of various nucleophiles. *Biol. Pharm. Bull. 24*(12), 1366–1369.

Jankowski, N. and K. Grabczewski (2006). *Learning machines*, pp. 29–64. Berlin: Springer-Verlag.

Jones, C. (2007). *Comprehensive glycoscience*, Chapter 4.31, pp. 569–605. Oxford: Elsevier Ltd.

Kamerling, J. et al. (2007). *Comprehensive glycoscience* (First ed.). Elsevier Ltd.

Kanehisa, M., M. Araki, S. Goto, M. Hattori, M. Hirakawa, M. Itoh, T. Katayama, S. Kawashima, S. Okuda, T. Tokimatsu, and Y. Yamanishi (2008). From genomics to chemical genomics: new developments in KEGG. *Nucl. Acids Res. 36*, D480–D484.

Kannagi, R. (2004). Molecular mechanism for cancer-associated induction of sialyl Lewis X and sialyl Lewis A expression – the Warburg effect revisited. *Glycoconj. J. 20*, 353–364.

Kannagi, R., Y. Fukushi, T. Tachikawa, A. Noda, S. Shin, K. Shigeta, N. Hiraiwa, Y. Fukuda, T. Inamoto, S. Hakomori, and H. Imura (1986). Quantitative and qualitative characterization of human cancer-associated serum glycoprotein antigens expressing fucosyl or sialyl-fucosyl type 2 chain polylactosamine. *Cancer Res. 46*, 2619–2626.

Kashima, H. and T. Koyanagi (2002). Kernels for semi-structured data. In *Proc. of the 19th ICML*, pp. 291–298.

Kawabata, S. and S. Iwanaga (1999). Role of lectins in the innate immunity of horseshoe crab. *Dev. Comp. Immunol. 23*, 391–400.

Kawano, S., K. Hashimoto, T. Miyama, S. Goto, and M. Kanehisa (2005). Prediction of glycan structures from gene expression data based on glycosyltransferase reactions. *Bioinformatics 21*, 3976–3982.

Kawasaki, T., H. Nakao, E. Takahashi, and T. Tominaga (2006). GlycoEpitope: the integrated Database of Carbohydrate Antigens and Antibodies. *Japan. Trend. Glycosci. and Glycotechnol. 18*(102), 267–272.

Keen, J. and A. Ashcroft (1999). *Post-translational processing*, pp. 1–42. Oxford University Press.

Kikuchi, N., A. Kameyama, S. Nakaya, H. Ito, T. Sato, T. Shikanai, Y. Takahashi, and H. Narimatsu (2005). The carbohydrate sequence markup lan-
guage (CabosML): an XML description of carbohydrate structures. *Bioinformatics 21*(8), 1717–1718.

Kikuchi, N. and H. Narimatsu (2003). Comparison of glycosyltransferase
families using the profile hidden Markov model. *Biochem. Biophys. Res.
Comm. 310*, 574–579.

Kilpatrick, D. (2002). Animal lectins: a historical introduction and overview.
Biochim. Biophys. Acta 1572, 187–197.

Kobata, A. (2007). *Glycoprotein Glycan Structures*, Chapter 1.02, pp. 39–71.
Elsevier.

Kojima, K., Y. Yamamoto, T. Irimura, T. Osawa, H. Ogawa, and I. Mat-
sumoto (1996). Characterization of carbohydrate-binding protein p33/41.
relation with annexin IV, molecular basis of the doublet forms (p33 and p41)
and modulation of the carbohydrate binding activity by phospholipids. *J.
Biol. Chem. 271*(13), 7679–7685.

Kornfeld, R. and S. Kornfeld (1985). Assembly of asparagine-linked oligosaccharides. *Annu. Rev. Biochem. 54*, 631–664.

Krambeck, F. and M. Betenbaugh (2005). A mathematical model of N-linked
glycosylation. *Biotechnology and Bioengineering 92*(6), 711–728.

Krengel, U. and A. Imberty (2007). *Lectins: Analytical Technologies*, pp.
15–50. Elsevier.

Krogh, A., M. Brown, I. Mian, K. Sjolander, and D. Haussler (1994). Hidden
Markov models in computational biology: Applications to protein modeling.
J. Mol. Bio. 235, 1501–1531.

Kuboyama, T., K. Hirata, K. Aoki-Kinoshita, H. Kashima, and H. Yasuda
(2006). A gram distribution kernel applied to glycan classification and
motif extraction. *Genome Informatics 17*(2), 25–34.

Lee, J., L. Baum, K. Moreman, and M. Pierce (2004). The X-lectins: a new
family with homology to the Xenopus laevis oocyte lectin XL-35. *Glycoconj.
J. 21*(8-9), 443–450.

Leffler, H. (2001). *Galectins structure and function - a synopsis*, pp. 57–83.
Berlin: Springer.

Leslie, C., E. Eskin, and W. Noble (2009). The spectrum kernel: a string
kernel for SVM protein classification. In *Pac. Symp. Biocompt.*, pp. 564–
575.

Ling, H. and A. D. Recklies (2004). The chitinase 3-like protein human cartilage glycoprotein 39 inhibits cellular responses to the inflammatory cy-

tokines interleukin-1 and tumour necrosis factor-alpha. *Biochem. J. 380*(3), 651–659.

Lohmann, K. and C.-W. von der Lieth (2003). GLYCO-FRAGMENT: A web tool to support the interpretation of mass spectra of complex carbohydrates. *Proteomics 3*, 2028–2035.

Lohmann, K. and C.-W. von der Lieth (2004). GlycoFragment and GlycoSearchMS: web tools to support the interpretation of mass spectra of complex carbohydrates. *Nucleic Acids Research 32*, W261–W266.

Loss, A., R. Stenutz, E. Schwarzer, and C.-W. von der Lieth (2006). GlyNest and CASPER: two independent approaches to estimate 1h and 13c nmr shifts of glycans available through a common web-interface. *Nucleic Acids Research 34*, W733–W737.

Lowe, J. and J. Marth (2003). A genetic approach to mammalian glycan function. *Annu. Rev. Biochem. 72*, 643–691.

Lutteke, T., A. Bohne-Lang, A. Loss, T. Goetz, M. Frank, and C. W. von der Lieth (2005). GLYCOSCIENCES.de: an Internet portal to support glycomics and glycobiology research. *Glycobiology 16*(5), 71R–81R.

Lutteke, T., M. Frank, and C.-W. von der Lieth (2004). Data mining the Protein Data Bank: automatic detection and assignment of carbohydrate structures. *Carbohydrate Research 339*, 1015–1020.

Lutteke, T., M. Frank, and C.-W. von der Lieth (2005). Carbohydrate Structure Suite (CSS): analysis of carbohydrate 3D structures derived from the PDB. *Nucleic Acids Research 33*, D242–246.

Lutteke, T. and C. von der Lieth (2004). pdb-care (PDB CArbohydrate REsidue check): a program to support annotation of complex carbohydrate structures in PDB files. *BMC Bioinformatics 5*, 69.

Marshall, A., C. Hendrickson, and G. Jackson (1998). Fourier transform ion cyclotron resonance mass spectrometry: a primer. *Mass Spectrometry Reviews 17*(1), 1–35.

Matsuno, H., Y. Tanaka, H. Aoshima, A. Doi, M. Matsui, and S. Miyano (2003). Biopathways representation and simulation on Hybrid Functional Petri Net. *In Silico Biology 3*(3), 389–404.

McLachlan, G. and D. Peel (2000). *Finite Mixture Models*. New York: John Wiley & Sons, Inc.

McNaught, A. and A. Wilkinson (1997). *Compendium of Chemical Terminology* (Second ed.). Blackwell Scientific Publications.

Merril, A. (2005). SphinGOMAP–a web-based biosynthetic pathway map of sphingolipids and glycosphingolipids. *Glycobiology 15*(6), 15G–16G.

Mironov, Jr., A., A. Luini, and A. Mironov (1998). A synthetic model of intra-Golgi traffic. *Faseb J. 12*, 249–252.

Mizuochi, T., T. Taniguchi, A. Shimizu, and A. Kobata (1982). Structural and numerical variations of the carbohydrate moiety of immunoglobulin g. *J. Immunol. 129*(5), 2016–2020.

Mizushima, T., T. Hirao, Y. Yoshida, S. Lee, T. Chiba, K. Iwai, Y. Yamaguchi, K. Kato, T. Tsukihara, and K. Tanaka (2004). Structural basis of sugar-recognizing ubiquitin ligase. *Nat. Struct. Mol. Biol. 11*(4), 365–370.

Mollenhauer, J. (1997). Annexins: what are they good for? *Cellular and Molecular Life Sciences 53*(6), 506–507.

Mortell, K., R. Weatherman, and L. Kiessling (1996). Recognition specificity of neoglycopolymers prepared by ring-opening metathesis polymerization. *J. Am. Chem. Soc. 118*, 2297–2298.

Narimatsu, H. (2004). Construction of a human glycogene library and comprehensive functional analysis. *Glycoconj. J. 21*, 17–24.

Needleman, S. and C. Wunsch (1970). A general method applicable to the search for similarities in the amino acid sequence of two proteins. *J. Mol. Biol. 48*(3), 443–453.

Nimmagadda, S., A. Basu, M. Eavenson, J. Han, M. Janik, R. Narra, K. Nimmagadda, A. Sharma, K. Kochut, J. Miller, and W. York (2008). Glyco-Vault: a bioinformatics infrastructure for glycan pathway visualization, analysis and modeling. In *Proc. Fifth Int'l Conf. on Information Technology: New Generations*, pp. 692–697.

Nitta, K. (2001). Leczyme. *Methods Enzymol. 341*, 368–374.

Packer, N., C.-W. von der Lieth, K. Aoki-Kinoshita, C. Lebrilla, J. Paulson, R. Raman, P. Rudd, R. Sasisekharan, N. Taniguchi, and W. York (2008). Frontiers in glycomics: bioinformatics and biomarkers in disease. An NIH white paper prepared from discussions by the focus groups at a workshop on the NIH campus, Bethesda, MD (September 11-13, 2006). *Proteomics 8*(1), 8–20.

Parodi, A. (2000). Protein glucosylation and its role in protein folding. *Ann. Rev. Biochem. 69*, 69–93.

Price, N. (2008). Permethylation linkage analysis techniques for residual carbohydrates. *Appl. Biochem. Biotechnol. 148*, 271–276.

Rabiner, L. (1989). A tutorial on hidden Markov models and selected applications in speech recognition. *Proc. IEEE 77*(2), 257–286.

Rabinovich, G., F. Liu, M. Hirashima, and A. Anderson (2007). An emerging role for galectins in tuning the immune response: lessons from experimental models of inflammatory disease, autoimmunity and cancer. *Scand. J. Immunol. 66*(2-3), 143–158.

Rabinovich, G., N. Rubinstein, and L. Fainboim (2002). Unlocking the secrets of galectins: a challenge at the frontier of glyco-immunology. *J. Leukoc. Biol. 71*, 741–752.

Raju, T., J. Briggs, S. Borge, and A. Jones (2000). Species-specific variation in glycosylation of IgG: evidence for the species-specific sialylation and branch-specific galatcosylation and importance for engineering recombinant glycoprotein therapeutics. *Glycobiology 10*(5), 477–486.

Raman, R., M. Venkataraman, S. Ramakrishnan, W. Lang, S. Raguram, and R. Sasisekharan (2006). Advancing glycomics: implementation strategies at the Consortium for Functional Glycomics. *Glycobiology 16*(5), 82R–90R.

Ranzinger, R., S. Herget, T. Wetter, and C.-W. von der Lieth (2008). GlycomeDB – integration of open-access carbohydrate structure databases. *BMC Bioinformatics 9*, 384.

Rao, V., P. Qasba, R. Chandrasekaran, and P. Balaji (1998). *Conformation of carbohydrates*. CRC Press.

R.K. Yu, M. Yanagisawa, T. A. (2007). *Glycosphingolipid Structures*, Chapter 1.03, pp. 73–122. Oxford: Elsevier Ltd.

Roberts, N., J. Brigham, B. Wu, J. Murphy, H. Volpin, D. Phillips, and M. Etzler (1999). A Nod factor-binding lectin is a member of a distinct class of apyrases that may be unique to the legumes. *Mol. Gen. Genet. 262*, 262–267.

Roy, R. (1996). Syntheses and some applications of chemically defined multivalent glycoconjugates. *Curr. Opin. Struct. Biol. 6*, 697–702.

Runza, V. L., W. Schwaeble, and D. N. Mannel (2008). Ficolins: Novel pattern recognition molecules of the innate immune response. *Immunobiology 213*(3-4), 297–306.

Sacchettini, J., L. Baum, and C. Brewer (2001). Multivalent protein-carbohydrate interactions. A new paradigm for supermolecular assembly and signal transduction. *Biochemistry 40*, 3009–3015.

Sahoo, S., C. Thomas, A. Sheth, C. Henson, and W. York (2005). GLYDE – an expressive XML standard for the representation of glycan structure. *Carbohydrate Research 340*(18), 2802–2807.

Sayle, R. and E. Milner-White (1994). RASMOL. Biomolecular graphics for all. *Trends Biochem. Sci. 20*, 374–376.

Schauer, R. (2000). Achievements and challenges of sialic acid research. *Glycoconj. J. 17*, 485–499.

Scholkopf, B. and A. Smola (2002). *Learning with kernels: support vector machines, regularization, optimization, and beyond.* Cambridge, MA: MIT Press.

Scholkopf, B., K. Tsuda, and J. Vert (2004). *Kernel methods in computational biology.* Cambridge, MA: MIT Press.

Sharon, N. and H. Lis (2007). *Lectins* (Second ed.). Springer.

Shawe-Taylor, J. and N. Cristianini (2004). *Kernel Methods for Pattern Analysis.* Cambridge University Press.

Smith, T. and M. Waterman (1981a). Comparison of biosequences. *Adv. Appl. Math. 2*, 482–489.

Smith, T. and M. Waterman (1981b). Identification of common molecular subsequences. *J. Mol. Biol. 147*(1), 195–197.

Smyth, P., D. Heckerman, and M. Jordan (1997). Probabilistic independence networks for hidden Markov probability models. *Neural Computation 9*(2), 227–269.

Stein, S., S. Heller, and D. Tchekhovskoi (2003). An open standard for chemical structure representation: the IUPAC chemical identifier. In *Proc. 2003 International Chemical Information Conference*, pp. 131–143.

Stevens, J., O. Blixt, L. Glaser, J. Taubenberger, P. Palese, J. Paulson, and I. Wilson (2006). Glycan microarray analysis of the hemagglutinins from modern and pandemic influenza viruses reveals different receptor specificities. *Journal of Molecular Biology 355*(5), 1143–1155.

Sud, M., E. Fahy, D. Cotter, A. Brown, E. Dennis, C. Glass, A. Merrill, Jr, R. Murphy, C. Raetz, D. Russell, and S. Subramaniam (2006). LMSD: LIPID MAPS structure database. *Nucleic Acids Res. 35*, D527–D532.

Suga, A., Y. Yamanishi, K. Hashimoto, S. Goto, and M. Kanehisa (2007). An improved scoring scheme for predicting glycan structures from gene expression data. *Genome Informatics 18*, 237–246.

Sugita, M., S. Itonori, F. Inagaki, and T. Hori (1989). Characterization of two glucuronic acid-containing glycosphingolipids in larvae of the green-bottle fly, Lucilia caesar. *J. Biol. Chem. 264*, 15028–15033.

Tang, H., Y. Mechref, and M. Novotny (2005). Automated interpretation of MS/MS spectra of oligosaccharides. *Bioinformatics 21*, i431–i439.

Tomer, K., M. A. Moseley, L. Deterding, and C. Parker (1994). Capillary liquid chromatography/mass spectrometry. *Mass Spectrometry Reviews 13*(5-6), 431–457.

Toukach, P., H. Joshi, R. Ranzinger, Y. Knirel, and C.-W. von der Lieth (2005). Sharing of worldwide distributed carbohydrate-related digital resources: online connection of the Bacterial Carbohydrate Structure DataBase and GLYCOSCIENCES.de. *IEEE Transactions on Knowledge and Data Engineering 17*(8), 1051–1064.

Tsai, C. (2007). *Biomacromolecules: introduction to structure, function, and informatics.* John Wiley and Sons, Ltd.

Umana, P. and J. Bailey (1997). A mathematical model of N-linked glycoform biosynthesis. *Biotechnol. Bioeng. 55*(6), 890–908.

Varki, A. and T. Angata (2006). Siglecs – the major subfamily of I-type lectins. *Glycobiology 16*(1), 1R–17R.

Varki, A., R. Cummings, J. Esko, H. Freeze, G. Hart, and M. Etzler (2008). *Essentials of Glycobiology* (Second ed.). New York: Cold Spring Harbor Laboratory Press.

Varki, A., R. Cummings, J. Esko, H. Freeze, G. Hart, and J. Marth (1999). *Essentials of Glycobiology.* New York: Cold Spring Harbor Laboratory Press.

Vasta, G., H. Ahmed, L. Amzel, and M. Bianchet (1997). Galectins from amphibian species: carbohydrate specificity, molecular structure and evolution. *Trends Glycosci. Glycotechnol. 9*, 131–143.

von der Lieth, C.-W., T. Kozar, and W. Hull (1997). A (critical) survery of modeling protocols used to explore the conformational space of oligosaccharides. *J. Mol. Struct. (Theochem.) 395-396*, 225–244.

Walther, D. (1997). WebMol – a Java-based PDB viewer. *Trends Biochem. Sci. 22*, 274–275.

Watanabe, K., E. Yasugi, and M. Oshima (2000). How to search the glycolipid data in LIPIDBANK for Web: the newly developed lipid database. *Japan. Trend. Glycosci. and Glycotechnol. 12*, 175–184.

Weis, W., M. Taylor, and K. Drickamer (1998). The C-type lectin superfamily in the immune system. *Immunological Reviews 163*, 19–34.

Xia, B., J. Royall, G. Damera, G.and Sachdev, and R. Cummings (2005). Altered O-glycosylation and sulfation of airway mucins associated with cystic fibrosis. *Glycobiology 15*(8), 747–775.

Yamanishi, Y., F. Bach, and J.-P. Vert (2007). Glycan classification with tree kernels. *Bioinformatics 23*(10), 1211–1216.

Yamanishi, Y., J.-P. Vert, and M. Kanehisa (2005). Supervised enzyme network inference from the integration of genomic data and chemical information. *Bioinformatics 21*, i468–i477.

Yang, R., G. Rabinovich, and F. Liu (2008). Galectins: structure, function and therapeutic potential. *Expert Rev. Mol. Med. 10*, e17.

Yoshida, Y., T. Chiba, F. Tokunaga, H. Kawasaki, K. Iwai, T. Suzuki, Y. Ito, K. Matsuoka, M. Yoshida, K. Tanaka, and T. Tai (2002). E3 ubiquitin ligase that recognizes sugar chains. *Nature 418*, 438–442.

Zelensky, A. and J. Greedy (2005). The C type lectin-like domain superfamily. *FEBS Journal 272*, 6179–6217.

Zhang, X. and M. Ali (2008). Ficolins: structure, function and associated diseases. *Adv. Exp. Med. Biol. 632*, 105–115.

Index